GKSS School of Environmental Research

Series editors: H. von Storch • E. Raschke • G. Flöser

Springer-Verlag Berlin Heidelberg GmbH

Hans von Storch • Götz Flöser (Eds.)

Models in Environmental Research

With 103 Figures

 Springer

VOLUME EDITORS:

Professor Dr. Hans von Storch
GKSS Research Centre
Max-Planck-Straße
21502 Geesthacht
Germany
E-mail: storch@gkss.de

Dr. Götz Flöser
GKSS Research Centre
Max-Planck-Straße
21502 Geesthacht
Germany
E-mail: floeser@gkss.de

ISSN 1437-028X

Library of Congress Cataloging-in-Publication Data
Models in environmental research / Hans von Storch, Götz Flöser, eds. p. cm. -- (GKSS School of
Environmental Research, ISSN 1437-028X) Includes bibliographical references and index.
ISBN 978-3-642-64028-5   ISBN 978-3-642-59563-9 (eBook)
DOI 10.1007/978-3-642-59563-9
1. Earch sciences--Simulation methods. I. Storch, H. v. (Hans von), 1949- II. Flöser, Götz, 1959- III.
GKSS School of Environmental Research (Series).
QE 33.2.M3 M64 2000   555'.1'1--dc21   00-046330

© Springer-Verlag Berlin Heidelberg 2001
Softcover reprint of the hardcover 1st edition 2001

Cover Design: Erich Kirchner, Heidelberg
Typesetting: Camera-ready by Ilona Liesner

SPIN: 10755681   32/3130/xz – 5 4 3 2 1 0 – Printed on acid free paper

Preface

Almost every scientific discipline uses "models". However, on closer inspection a large variety of different approaches are labeled as "models". For some scientists models are per se complex numerical models implemented on computers, but for other disciplines models take the form of mechanical analogs. Still others consider "models" to be preforms of theories or to be restricted concepts, within which basic dynamical (natural science as well as social science) aspects can be understood and described. Some models are not built in a systematic scientific manner, but evolve through social processes, and become powerful agents in the political arena. Thus, the applications of models range from conceptualisations, illustration to guidance of the decision process in everyday life and world politics.

Thus, models are a key concept in sciences, and specifically so in environmental sciences, where experiments can hardly be conducted because of the open and non-replicable character of environmental systems. Much too often modeling is seen and taught as a mere technical process of differential equations, numerics and computation, while the philosophical implications (the role of models in the process of generating knowledge) are not really thought about. This is a rather unfortunate limitation, as it prevents scientists from both acknowledging the inherent limitation of models and their applicability and fully exploiting their potential in providing added value to observations and theories. Therefore the topic "Models in Environmental Research" was chosen for the Second GKSS School of Environmental Research, which was held on 23-30 September 1998 in the Zündholzfabrik on the bank of the river Elbe in Lauenburg near Hamburg. As with the First School "Anthropogenic Climate Change" (von Storch, H., and G. Flöser (Eds.), 1999: Anthropogenic Climate Change. Proceedings of the First GKSS School on Environmental Research, Springer Verlag, ISBN 3-540-65033-4) about 50 international graduate students and ten lecturers discussed for one week various aspects of modeling from a broad range of viewpoints.

The lectures of the Second GKSS School of Environmental Research "Models in Environmental Research" are collected in this volume. The philosophy of modeling from the standpoints of a sociologist and of a natural scientist are discussed in the first two chapters. The physical and mathematical principles in constructing dynamical models are presented in Chapters 3 to 6, covering examples from oceanography and meteorology. In contrast, different modeling concepts in ecology and material research as well as morphology are offered in Chapters 7 to 9, while the use and role of statistical techniques is presented in the last two Chapters 10 and 11.

Before acknowledging the help of organisations and people in setting up the school, a few words about GKSS, its environmental research activities and the school may be in order: GKSS is a member of the Hermann von Helmholtz Gemeinschaft Deutscher Forschungszentren (HGF). Part of its research addresses environmental problems with emphasis on water and climate in the coastal zone. Its main interests are related to regional climatology and climate change, with interdecadal variations in the state of the coastal oceans and the flow of heavy metals, nutrients and other materials in river catchments to the coastal zones. This research aims at the understanding of changes in the environment, both because of internal (natural) dynamics and because of anthropogenic interference. In order to disseminate results of this research as well as to initiate a broad discussion among senior scientists in the field and younger colleagues from all over the world, the Division of Environmental Research at GKSS has instituted the GKSS School on Environmental Research. The present volume covers the topics of the second school; the first school, organised in 1997, dealt with anthropogenic climate change. A third school, on reconstruction paleoclimatic states, is presently being prepared in cooperation with other HGF institutions. Applied environmental research has always been aware of the societal implications and boundary conditions. Therefore, every discussion about environmental change should comprise a social component; in our school we acknowledge this need by having social scientists among our lecturers.

While the second school was funded mainly by the GKSS Research Centre, it received significant financial assistance through the Verein der Freunde und Förderer der GKSS, Carl Duisberg Gesellschaft, University of Hamburg, and the German Climate Computing Centre. The technical and administrative support of Ilona Liesner and Beate Gardeike is gratefully acknowledged.

Hans von Storch, Lauenburg, January 2001

Contents

Chapter 11 Statistics – an Indispensable Tool in Dynamical Modeling.....203
by Hans von Storch

List of contributors

Aike Beckmann
Alfred-Wegener-Institute for Polar and Marine Research, Postfach 12 01 61, D-27515 Bremerhaven, Germany

Karl-Heinz van Bernem
Institute of Hydrophysics, GKSS Research Center, Max-Planck-Strasse, D-21502 Geesthacht

Wolfgang Brocks
Institute of Material Research, GKSS Research Center, Max-Planck-Strasse, D-21502 Geesthacht

Heike Langenberg
Nature (Editorial), Porters South, 4-6 Crinan Street, London, N1 9XW, Great Britain

Peter Müller
Department of Oceanography, University of Hawaii, 1000 Pope Road, Honolulu, HI 96822, U.S.A.

Michael Schatzmann
Institute of Meteorology, University of Hamburg, Bundesstraße 55, D-20146 Hamburg

Michel Schmitt
École Nationale Supérieure des Mines de Paris, Centre de Géostatistique, 35, rue Saint Honoré, F-77305 Fontainebleau-Cedex, France

Nico Stehr
Sociology, University of Duisburg, Lotharstraße 65, D-47057 Duisburg, Germany

Hans von Storch
Institute of Hydrophysics, GKSS Research Center, Max-Planck-Strasse, D-21502 Geesthacht, Germany

Hans Wackernagel
Centre de Géostatistique, Ecole des Mines de Paris, 35, rue Saint-Honoré, F-77305 Fontainebleau-Cedex, France

Chapter 1
Models as Focusing Tools:
Linking Nature and the Social World

by Nico Stehr

Abstract

My paper explores the nature of models as focusing devices in the sciences in general, and the symmetries and asymmetries found in the prevailing global change models linking the relationships between physical processes and social milieus in particular. The asymmetries are linked not merely to the subordinate status assigned to social sciences in these models: either a social science model is seen as providing information about human activities that perturb natural processes or the model takes its biological or physical outcomes as given and attempts to work out the economic, social and political consequences. An argument is made that social science considerations ought to be located in different types of models of global change processes.

1.1 Models as Focusing Tools: Linking Nature and the Social World

> Owing to the contemporary mania for what are called facts, we are apt to forget that an age can only learn to know itself if the different methods of approach, the power of formulation, and the analysis of complex phenomena do not lag behind the collection of data. It is not enough that our age should be rich in a knowledge of fundamental facts, which gives it ample scope for new experiences; it must also frame its questions adequately. This it can only do if the tradition of theoretical formulation is held in the same esteem as the technique of sheer fact finding.
> Karl Mannheim

> It is easier to draw schematics than to describe what actually occurs.
> Kates 1985:14

Although the term model is extremely vague, the construction of models is one of the flourishing intellectual enterprises in science. Any discussion of the role and the functions of models in the natural and social sciences invariably gets entangled in highly contentious philosophical debates about such matters as the status of language, reality, explanation, truth, data, understanding, description, constructivism, theory, and so on. A discussion of models in science that makes adequate reference to all these matters and how they have been linked in various traditions in the philosophy of science would be an undertaking without any readily apparent closure. Moreover, models are all over the place: Take the entry under the heading "model" in a major encyclopedia (The American Heritage Dictionary of the English Language. New York: Houghton Mifflin 1969):

1. A small object, usually built to scale, that represents some existing object.
2. A preliminary pattern representing an item not yet constructed, and serving as the plan from which the finished work, usually larger, will be produced.
3. A tentative ideational structure used as a testing device: "two conflicting models of generative grammar" (Noam Chomsky).
4. A style or design of an item: His car is last year's model.
5. A person or object serving as an example to be imitated or compared: "in her temper, manners, mind, a model of female excellence" (Jane Austen).
6. A person or object serving as the subject for an artist or photographer.
7. A person employed to display clothing by wearing it.

Since the list is by no means complete; let me add the following normative considerations:

1. Models are good things.
2. Mathematical models are even better.

The dilemma is quite real. However, the multiplicity of meanings is self-exemplifying. It has the merit that offers a first hint, perhaps even quite a significant answer to the question of the role or function of models in science. That is to say, as one strives to generate a model of models, we realise that one of the main functions if not the primary function of models is to reduce the complexity of domains of inquiry. But what model of inquiry into models does one employ in the context of asking about the role of models in science? Where do we start, what are some of the important questions, what answers are we looking for, and how do we judge any answers we may find?

My approach to the question of the role of models in science is best described as pragmatic. My observations about models as focusing tools do not try to conform to one of the various prescriptive traditions in the philosophy of science and what they may reveal about the "real" function of models. I am exploring what ways model can help us to generate insights into the workings of nature and society.

A pragmatic approach is justified by the observation that the term model enjoys a broad range of meaning and uses in the sciences. It refers as we have seen to anything from a physical object in a display case to an abstract set of ideas. It would be foolish to believe that it is possible to reduce the multiplicity of meanings of the concept in order to arrive at certain consensual or even essential features of models.

If I may be permitted to offer one rather general observation about the practice of modeling in spite of my own expressed misgivings about making generalisations in this context, I think it would be fair to say that natural scientists prefer as their formal models models-in-use, perhaps even models that employ formal language. The models that circulate in the social sciences are much more likely to be analogies and metaphors. One should not *a priori* belittle the use of metaphors and analogies as impotent cognitive devices (see Weingart and Maasen 1997). They can be powerful focusing devices if measured by the degree to which they generate advances in knowledge. Donald Schon 1963 has alerted us to the process and the importance of the "displacement of concepts" and the central role this process may play in an understanding of the conditions for the very possibility of theoretical innovation and imagination.

One of the major controversies in the philosophy of science extends to the question as to whether models in science are mere aids to theory construction that can be discarded once a "theory" has been developed, a position the French physicist Pierre Duhem ([1914] 1954) advocated; or whether the roles of models go much beyond their function as aids, as the English physicist Campbell in response to Duhem argues. For Campbell (1920:129), models are "an utterly essential part of theories, without which theories would be completely valueless and unworthy of the name".[1]

My observations about models as focusing are divided into two parts. In addition to a few remarks about the practice of modeling, I will have some observations about the methodology of modeling. Not surprisingly, there is not a common set of methods. However, two attributes that are frequently mentioned in such discussions; they concern (1) the quality of isomorphism of models and (2) the issue of quantification. In the second, substantive part of the lecture, I will propose a model designed to offer some insights into what I call *societal sensitivity*. The metaphor social sensitivity is, in a sense, an example of the process of cognitive displacement, and therefore of the cognitive work models may be able to accomplish.

1.2 The Practice of Modeling

Let us look at the ways in which models do operate in the scientific community. Judging from models-in-use, what we are able to say is that models in the social and natural sciences function as "focusing devices". Models serve heuristic purposes in science.[2] Models refer to a simplified representation of what is thought[3] to

---

[1] See Mary Hesse's (1966) extensive discussion of and effort to move beyond these two positions.

[2] And in this sense, the term "model" indeed has affinities to a variety of notions, for example, the concept of "ideal types" as proposed by Max Weber ([1921] 1972: 3,4,10) or the concept of models as "works of fiction" (cf. Cartwright, 1983:153) that all serve the cognitive function of simultaneously focusing attention towards some and away from other possible features of a phenomenon.

[3] Achinstein (1965:102-105) has systematised what h considers to be the most important elements of models in science as follows: "1. A theoretical model consists of a set of assumptions about some object or system. 2. A theoretical model describes a set of exhibited properties of an object and its inner structure. 3. A theoretical model is treated as an approximation useful for certain purposes. 4. A theoretical model is often formulated, developed, and even names on the basis of an analogy between the object or system described in the model and some different object or system."

be an underlying more complex reality. Models are generated around certain constitutive elements and relations between elements. A critique of modeling should not, as a result, be a lament that models mainly distort reality. Models perform the essential, unavoidable cognitive task of selecting an aspect of reality for attention. Thus, one cannot criticise the use of models per se. What can be critically examined, however, are the grounds for selecting specific elements for attention and the postulated relationships among elements. The exact function and aspirations connected with model-in-use is context-dependent. Models are constructed with specific purposes in mind.

A model is fashioned with the aim of bringing forward and alerting us to some aspect of the world, while leaving others in obscurity or even darkness. For example, proposing that gas particles are akin to a collection of billiard balls in motion is meant to alert us to the possibility that there are some properties of molecules that are analogous to billiard balls: for instance, molecules exert no force on each other except on impact or travel in straight lines except at instants of collision (see Hesse, 1966:8). The reference to billiard balls brings forward features that are very much unlike those that we would perhaps invoke if we proposed it is not billiard balls but football players in motion that behave like gas particles.

Models as focusing devices inevitably have both certain strengths and considerable weaknesses. This already follows from the observation that a way of seeing is also always a way of not seeing. In the context of standard economic discourse, the dominant theoretical models expose us to rational conduct, scarcity, allocation and exchange in a steady state and cut off from the environment. Undoubtedly, the focus of the model enforces and highlights some features of economic, social and cultural reality but inhibits us from seeing other processes that affect economic conduct (see Lundvall 1992). Another prominent example from the social sciences, Max Weber's widely cited ideal type of bureaucracy, deliberately attends to what bureaucratic structures attain: precision, reliability, efficiency. Needless to say that another way of seeing would stress the opposite virtues and dysfunctions of bureaucracies.

1.3 The Methodology of Modeling

The methodology of modeling is at best a form of tacit knowledge that is rarely made explicit and succeeds in generally restricting ways in which models are constructed and employed. But there are no generally accepted rules, much less laws of inference in model building or precise decision-making procedures. Model building is analogous to ways of searching. It would appear, therefore, that interpretative skills, imagination, hunches etc. are the main "methods" that drive model building. Nonetheless, in the discussion of models much emphasis is placed on the correspondence between the model and reality or the quality of isomorphism of models, as well as the topic of the quantification of models.

1.4 Isomorphism

It is often assumed that there are many isomorphic models in the natural sciences while there are few if any in the social sciences; unable to "construct isomorphic models, the social sciences most rely more heavily on other types of models and on other analytical techniques" (Gordon 1991:108). Moreover, there is some agreement that models representing the complexity of the world more closely are superior to models that fail to live up to the same standard. Models that represent reality more fully are assumed to generate knowledge that is of greater precision, detachment, object centeredness, certainty, and practical use. Referring to but one of the enumerated virtues of reality conforming models, it is almost taken for granted that policy advice as well policy based on such models will be more effective in the end (cf. Stehr 1992).

Assuming these observations are generally correct, the questions that follow are (1) what exactly is the nature of the distance between models and reality, and (2) is the intellectual or practical utility of models linked to their degree of isomorphism?

The first issue arises simply because many modelers as well as critics of models tend to assume – without warrant I may add – that reality and models *can only be seen to vary independently* of each other and that the central issue therefore becomes how close or how distant reality and its representation are or how isomorphic models in fact are. This is an assumption that is not always justified because it is rather silent, for example, on the issue of self-fulfilling or self-defeating prophecies as well as unintended consequences of social (cognitive) action. That is to say, a model, though isomorphic, may well cease to have such a quality not because its "creators" make the "required" adjustments to bring it in line with reality but because reality adjusts so to speak to the model and succeeds in realising its construction. While such a reservation applies with obvious force to social phenomena, it has some bearing for natural processes as well, since efforts to bridge models and reality are driven by social considerations. After all, nature does not talk to us directly.

With respect to the second issue, whether models that are reality-congruent are more useful in practice and whether models that are isomorphic intellectually are more desirable; there are advocates of modeling that would be prepared to argue that the usefulness of models within scientific discourse is not driven by their "realism": "The real test of a model ... is whether is works effectively as a scientific instrument, not the degree to which it replicates the real world (Gordon, 1991:108). By the same token, the great majority of perspective in the philosophy of science concerned with the question of the usefulness of scientific knowledge in practice would maintain that it is on the basis of reality-congruent models that useful knowledge is generated.

1.5 Quantification

It appears to be an almost taken for granted assumption that the conceptual elements that form a part of models must be quantifiable attributes. The conviction is widespread that quantitative assertions and models are not only the necessary but also the sufficient condition for a subject matter to be scientific. The greater usage of models in economics that are quantified therefore is seen not only as an expression of the relative ease with which economic conceptual attributes are quantified also but testimony to the greater scientificity of economic discourse when compared to other social science disciplines. In such a context, as a result, discussions about the virtue of qualitative assertions, and about the significance of metaphors and analogies are difficult to sustain.

1.6 Modeling Societal Sensitivity

The substantive focus of my paper is on a critical analysis of the nature of the linkage construed in models designed to capture the interaction between nature and society, more specifically, the linkage between climate and society.[4]

We are faced today with a number of incongruities in science and society: On the one hand, there is the increasing transformation of the world by science and the immense growth in scientific activity; on the other hand, there are, often as a more or less direct outcome of the former, both new classes of social and environmental threats and challenges that give rise to issues and challenges, as well as the question of systematic cognitive uncertainty (cf. Funtowicz and Ravetz, 1990). What is evident, however, is that mere expansion along the traditional trajectory of the development of science, namely greater and greater specialisation or the conventional hope for a growing control of nature by science and greater certainty achieved by scientific knowledge cannot hope to address these challenges effectively.

There has neither been a comprehensive study of the problems of interdisciplinary research nor has there been a thorough examination of the methodological issues of an integration of scientific apples and oranges. But closer to the specific points at issue here, most discussions of climate change, climatic impact and climate policy are carried out in terms of models that rely on enormous and continuously expanding intellectual resources that have been spent since the early 1960s on efforts to estimate "climate sensitivity" in terms of its response to a doubling in

---

[4] My focus here is not on the technical "nature" of climate modeling as practiced today (cf. Shine and Henderson-Sellers, 1983) or whether climate models can be verified at all (cf. Oreskes et al., 1994). But the typical output of climate models or the dependent variable (for instance, numerical predictions about central meteorological dimension such as temperature or precipitation) are the subjects of the analysis. In other words, we are not primarily interested in the nature of General Circulation Models (GCMs) as such insofar as these models of the climate system tend to be restricted to the physical world; however, the output of GCMs is of interest since such outputs (for example, how regionally specific are the outputs) invariably link such models in specific ways to the social world. Such outputs are seen to constitute the concrete linkage between climate model, society and policy.

the concentration of atmospheric CO_2.[5] But limited, if any, cognitive energy and resources have been dispensed on research of the societal sensitivity of climate.[6]

The examination of the models constructed to capture environmental processes is significant because the problems they construe for society to tackle derive not from the direct personal experience or everyday life but are recognised and constructed scientifically. Without the information provided by scientists about ozone depletion, who in society could or would have claimed to perceive such an issue? The role of scientific expertise and the trust experts enjoy in society plays a significant role (cf. Stehr, 1994). In addition, the models are said to touch upon fundamental social and political questions; for example, issues such as national security, cultural identity, international political and economic relations, social inequalities and perhaps even the survival of the humankind. As a result, it is doubtful that conventional methods and techniques such as cost-benefit analysis[7] or the existing division of labor within science will be sufficient to develop new and effective knowledge.

1.7 Examples of Models

Within the social sciences, economics appears to be the discipline that most frequently and self-consciously deploys (formal) models. In one of the first models of this kind, exemplified by economic ideas of the Physiocrats, (that is, by their "tableau economique", Fig. 1.1), economic activity is represented as a flow of expenditures between the three classes in society.

Many of the assessment reports for policy makers that have been drafted in recent years and that deal with anthropogenic climate change also reflect the fact that economics among the social sciences has the greatest intellectual affinity to more formal and systemic efforts in modeling. However, the models found in climate research designed to depict the interaction between climate change and a broad range of economic and social activities usually are not developed to describe the structure of empirical observations (data) of past occurrences; rather, they are developed to predict events and phenomena.

1.8 The Poverty of Economics

Insofar as economic theory is the driving force of most, if not all social science driven models of global change, the question has to be answered: how good is contemporary economic theory? I do not want to add to the lament mostly by

---

[5] Since 1979, the quantitative estimates for climate sensitivity which constitute the most important output of climate models for assessment efforts have remained constant to the present (see van der Sluijs et al. 1998:2993).

[6] Efforts that attend to the societal sensitivity often have been informed by the ill-fated tradition of climate determinism.

[7] Strong advocates among economists of the utility of cost-benefit analyses in environmental issues and policy advice are Nordhaus (1991), Barbier and Pearce (1990).

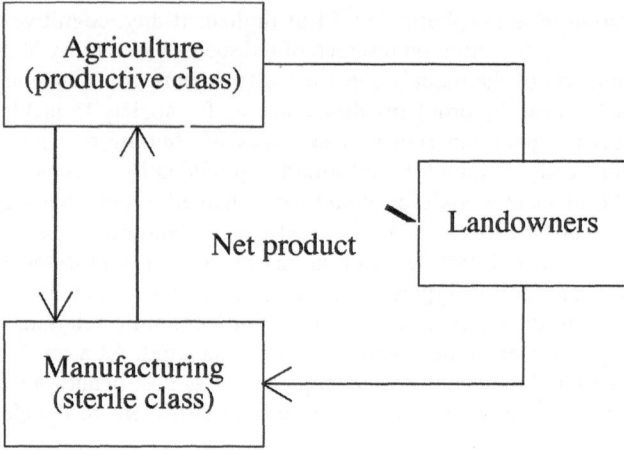

Fig. 1.1. Tableau economique

non-economists, of course) about the narrow conception of *Homo Economicus* but focus instead on the practical efficacy of contemporary economic reasoning.

One does not necessarily have to share the wholesale indictment of macroeconomic theory as irrelevant to economic policy because it is incapable of anticipating the course of economic affairs (e.g., Drucker 1993:95-99); however, it is evident that economic theory compared to past successes of economics fails to generate effective policies, since some of its central assumptions are no longer isomorphic with economic and political reality (cf. Stehr 1994). At the same time, economic discourse subscribes to the general prohibition in social science discourse that socio-economic and natural processes ought to be clearly separated.

The main logic employed in modeling efforts designed to assess the consequences of global climate change has been economic reasoning, in particular some type of cost-benefit analysis. The sole objective of cost-benefit analysis is the goal of economic efficiency. It follows that any increase in total net benefits or avoidance of net costs is considered desirable, irrespective of the consequences such action may have on the distribution of these benefits or costs. By the same token, consideration of other especially non-economic objectives, as well as costs is considered not to be part of a typical cost-benefit analysis.[8]

But what are some of the possibly more inclusive ways of modeling the societal sensitivity to climate? I will suggest that it might also be helpful in an attempt to overcome the separation between allegedly pure categories of social and natural to think of the relation between physical and societal processes as composite or hy-

---

[8] Rees and Wackernagel (1994:363-364) have developed the notion of and attempted to measure the "natural capital requirements of the human economy" that explicitly links social and natural order expressed as the "carrying capacity" in terms of the maximum population of an organism a habitat can support indefinitely. They advocate the need for such a measure last but not least because "orthodox economic analysis is so abstracted from biophysical reality that its ability to detect, let alone advise on critical dimensions of carrying capacity, is severely compromised".

brid phenomena; that is, of emergent processes in which both social and physical activities are imprinted. The interaction results in new, namely hybrid processes in which the distinction between that which is social and what is natural disappears.

1.9 Hybrid Forms or the Linkage between Social and Physical Processes

In contrast to such an approach favoured by economists when modeling social and physical processes, the much more common social science approach continues to be based on a clear separation of social from natural and natural from social; as well as work on each set of these phenomena along the established lines of the division of labor in science.

The more recent claim that society and nature are indivisible, or the observation that nature and society are not separate but - once again - have to be seen as intertwined becomes a more and more common and, as is argued, urgent assertion among social scientists. Nonetheless, the absence of nature in social science discourse is common in social science today. It is also one of the great success stories of modern society and modern social science. As a kind of cultural accomplishment nature has been dropped from both. Thus, social science is better understood as the separation and the reasons for the divorce of society and nature.

I would like to propose that one might be able to develop the idea of *societal sensitivity* in analogy to the idea of climate sensitivity in an effort to overcome the difficulties now experienced in theories that attend the impact of nature on society.

The notion of *climate sensitivity* has different meanings to different sets of actors involved in the climate debate. But in the last twenty years, within the climate science community, a remarkable and robust consensus has emerged with respect to the construction of estimates of climate sensitivity. The estimate and therefore the practical output of climate models is typically expressed in the following manner: "We estimate the most probable global warming for a doubling of CO_2 to be near $3°$ C with a probable error of $±1.5°$." (US National Academy of Science, 1979: note 2, 2).

I would like to interpret the notion of climate sensitivity to refer to the imprint of social order in natural order. In analogy to this notion, societal sensitivity refers to the imprint of natural order on social order, and that such an imprint cannot be captured by mere economic reasoning.

However, it ought to be clear from the outset that the output of any models of the societal sensitivity to climate change will never be that terse or, for that matter, robust and reach a similar degree of concurrence in the scientific community.

The answer to the question of societal sensitivity to climate would come quite easily if I would be asked to respond to this question from either the point of view of a contemporary historian sociologist who has examined the influence of climate on the emergence and the nature of modern society, or if I would formulate my answer on the basis of the perspective of geographical discourse during the first part of this century.

The French historian Le Roy Ladurie ([1967] 1988:119), for example, observes that "in the long term" the human consequences of climate appear "to be slight, perhaps negligible, and certainly difficult to detect." The narrowness of the range of secular variations "and the autonomy of human phenomena which coincide with them in time, make it impossible for the present to conclude," as Le Roy Ladurie ([1967] 1988:275) stresses in his study of the interaction between climate and history since the year 1000 that "there is any causal link between them." That is to say that today, the American geographer Ellsworth Huntington's (1927:138) tireless work at the beginning of this century affirming the thesis that "climate paints the fundamental colors on the human canvas" appears to be amusing to some, to others extreme or lazy (Le Roy Ladurie [1967] 1988:24), but most would likely consider it absurd and therefore certainly on the very margins of social science discourse in regard to the impact of environmental factors on the human condition. Although the discussion of the impact of climate on societies did not cease abruptly in social science, it ultimately was discredited, and only fairly recently, vanished almost without any trace as a largely compromised and widely discredited line of inquiry. It therefore has become more common today to find it "amusing to think that the men of former times would not have been put out by ...climatic explanation, implicating as it does the heavens" (Braudel [1979] 1992:51).

But how do we move from Huntington's naively realistic and mechanistic idea that climate "works" to the less fatalistic and much more open notion that climate "matters?" Such a model of societal sensitivity to climate requires the firm refusal to succumb to the seductive simplicity of climatic determinism and fatalistic utopias as well the notion that climate is a mere social construct.

In a most general sense, I want to propose that natural and social processes are mainly imprinted into the *boundary conditions* of nature and society. In the case of society I suggest that although it is significantly shaped by historical or selective constructions, our understanding of and encounter with climate is in important ways affected by, resonates with, and is shaped by its "extreme" responses – that at times may well be the consequence of human interventions into global climate processes.[9] That is, extremes constitute a form of feedback by nature that is read by or absorbed into society.[10] This is by no means the only well known fascination and even rapture with extremes of all descriptions among the public and the media in our times. Varied, even "ritualistic" cultural responses to extremes ultimately display and often celebrate the familiar, namely, "normal" patterns and homeostatic processes. Extremes constitute and are apprehended as a temporarily dis-

---

[9] This assertion about the significance of extremes should not be misunderstand to represent a kind of ontological thesis rejecting any kind of "gradualism" (for example in natural history, see Gould 1980:226) but as an empirical hypothesis about the practical ways in which nature becomes pertinent for society.

[10] Although our example refers to climate and its possible imprint on society, the importance of extremes as issued in nature of course is not limited, we would contend, to climatic phenomena. As a matter of fact the literature in various fields stresses the salience of extremes for society; for example, Robert E. Park (1936:5) in his classical discussion of the "nature" of human ecology repeatedly refers to the significance of extremes (such as famines, epidemics) in producing "disequilibria" and institutionalised responses that slowly but surely obliterate the extreme as the triggering event. Park specifically refers in some detail to the advent of the boll weevil disease in the early 19th Century and the effects it had for agriculture in the Southern United States.

turbed equilibrium. And it is in this sense that climate extremes violate taken-for-granted and trusted conceptions and observations about climate (Stehr 1997). Although they are not static over time, what is experienced as climate extremes are anomalies and disappointments. Climate extremes remind us of the reality hidden behind the surface of the climate construct. Climate extremes offer and manifest the resistance of natural reality in the background of the social climate construct. They allow for the possibility of observing and criticising our observations about climate. In order to observe our observations about climate or to be reminded of the effects of climate on society, we have to step back and be forced to leave accepted constructs that obliterate climatic forces. Climatic extremes matter as a mechanism that precisely accomplishes this feat.

In the past, society responded to climate extremes by imprinting them into social action and it continues to do so. Climate *extremes* are institutionalised in society, they are inscribed in the forms of myths, ideologies, stories (including more or less elaborated narratives of nature in everyday life), technologies, regulations, and organisations. An obvious as well as stable and powerful example of this institutionalisation are protective dikes erected at both rivers and oceans as well as the laws and regulations that govern their construction, maintenance and use. In much the same way, the evolution of shelter, clothing and nutrition is to some extent an inscription of climate extremes into the social fabric (cf. Latour 1991). Climatic extremes are engraved and objectified in the construction, maintenance and utilisation of many of the modern means of transportation. Modern instruments of transportation are not only utilised to link open spaces with each other and carry commodities, information and humans but they constitute artifacts that are responses to climate, especially climatic extremes (for approaches to Large Technical Systems, see Mayntz / Hughes 1988; Summerton 1994). In a way, means of transportation are portraits of and embody social encounters with climate. Of course, importantly such encounters manifest themselves in efforts to exclude or, to draw boundaries of exclusion for climatic extremes. Transportation takes place in familiar spaces, artificial zones and fairly tight enclosures that keep out undesirable climatic conditions. None the less, engraved into the enclosures that we inhabit (or, in which we move) are climatic conditions or, if you like, nature that is not "our" nature but nature from which we desire to withdraw. The greater the distance such artifacts have to travel, the greater the likelihood that climatic extremes are inscribed into the construction of the object. As time and distance become increasingly irrelevant to social and economic life, the greater the influence of climatic extremes on the design of such artifacts. Paradoxically, as these extremes are built into the object, they tend to vanish from view and certainly from direct experience and encounters.[11]

Although nature manifest in climate processes may be institutionalised in society and indeed take on moral qualities (as in the trope "nature strikes back"), the institutionalisation of nature paradoxically converts climate into an almost invisible entity. As the obvious it becomes hidden. The institutionalisation of climate in

---

[11] Cf. the following criticism of Huntington made by Bates (1952:120): "The western European environment, lauded by Huntington and his followers as ideal for the development of civilisation, was an insurmountable obstacle to civilisation until methods had been found for mitigating its effects."

society paradoxically means to distance society from climate and to decrease the contingencies for society that may arise from climate. The successful reduction of the contingencies that derive from climate allows for an increase in the contingencies that come with the socio-cultural development of knowledge. Here the argument comes full circle: it is now no longer nature which poses threats, but it is uncertain knowledge which troubles us.

Appendix: Additional Comments of Modeling Climate and Societal Sensitivity

Most of the effects that climate models describe as resulting from human activity have to do with the energy balance and ultimately the temperature distribution. There are many ways in which human action may affect the heat balance of the globe. Some of these activities are by no means actions confined to recent times. Humans have altered vegetation patterns for centuries, even before cultures had developed permanent settlements. What is at issue specifically is of course the burning of carbon and **release** of carbon dioxide into atmosphere. The volume of CO_2 in parts per million by volume (ppmv) is said to have risen from an estimated 280-290 at the beginning of the industrial revolution to 365, at the present time and may double by in another half century (cf. Kellogg 1978:210–211).

One of the unavoidable asymmetries in the models is that carbon dioxide represents a *global condition* while many of the consequences and responses (policy implications) are regional and local.

Similarly, the unfolding of the temperatures increments over time are relatively precise while the "extension and enlargement" of social action (cf. Stehr 1994:29-32) generally and in regards to climatic changes (assuming that such transformations can be disaggregated in this fashion), are complete unknowns.

The specific connection between climate change, conceptualised as a physical process in general circulation models and social activities in available models is most often represented by changes in weighted average global surface temperature over specific periods in time rather than variability, extremes, seasonal cycles or any of the other meteorological dimensions.[12] As Kellogg (1978:216-217) notes: "The rate of change of the temperature is probably the most important factor to consider, *since it can be compared to the statistics of natural climatic change*. At issue is primarily an increment over a certain time-span in one aspect of the typical climatic conditions human face. The change is due to human activity and may or may not be superimposed on a warming or cooling trend due to natural changes.

The naming of the problem affects the ways in which the issue may be understood; for example, "global warming", the "greenhouse effect" etc. signal that the issue at hand is an increase in temperature and that the most relevant solution is one that attacks the agent that causes it.

---

[12] A first critique of the dominant output and the dubious value of such models for climate policy may be found in Katz and Brown (1992).

While the focus is almost exclusively on temperature increments, a multiplier effect of enormous proportions may be observed once, (as is virtually always the case) hypotheses about induced changes in other meteorological attributes such as precipitation are advanced as well. As a result, consequences and choices suddenly multiply. Meteorological dimensions other than temperature or, to be more precisely, ranges and limits of temperature may be much more significant for social activities, because there are no effective and/or economic means to shield social activities from such dimensions. Precipitation patterns for example may have a much more salient impact on social action. They effect, in a very real sense, "where deserts, marginal lands and 'food baskets' will be" (Kellogg 1978:219). One cannot claim at this point that we have very good knowledge about the relative social significance of different meteorological attributes and combinations of attributes.

There is an inevitable tendency in discussion of the implications of these models to define *and* seek solutions by referring to the agent of change, namely "fossil fuels" or carbon dioxide (cf. Schelling 1983:449).

Most of the terminology employed to conceptualise and describe the exact interrelations between physical processes and social milieus in modeling exercises tend to be extremely ambivalent if not entirely empty of any detail. As a result, it is justified to refer to the linkages typically invoked in these models as black boxes.

Among these virtually empty assertions is most importantly the fundamental thesis that a warming will produce severe damages (or blessings for that matter, cf. Stehr 1995) for social activities. As such, this assertion itself is not immediately persuasive. It is not immediately convincing, because millions and millions for tourists for example spend billions to travel to "climates" in search of warmth.

The modeling efforts do not extend to and therefore also justify the term black box models; a *theory of society* explicates and answers general questions about what the social world will look like in a few decades from now. Unless one implicitly reverts to the long discredited mono-causal climate theories, that were much in vogue during the 18th century, climate models do not advance any credible notions about the ways in which people will live and about developments that will affect their existence.

It would also be misleading to superimpose gradual temperature increments on society as we know it today; for example on today's agriculture, economy and structure of inequality, today's patterns of existence and political map, its urban complexes, technologies, the labour market, consumption patterns and values, living standards, diet, or leisure activities (cf. Schelling 1983:453).[13] Unless we

---

[13] Mendelsohn, Nordhaus, and Shaw (1994:755) have examined the impact of global warming on land prices. Aside from their findings that American agriculture may actually derive economic benefit from global warming, if one employs the crop-revenue rather than the cropland model, their analysis also indicates that the assumption of static *or unchanging patterns of farming activity* (the dumb farmer scenario) can be misleading: "The production-function approach will overestimate the damages from climate change because it does not, indeed cannot, take into account the infinite variety of substitutions, adaptations, and old and new activities that may displace no-longer advantageous activities as climate changes." But there is not only the dumb farmer scenario with respect to a changing climate, there is also the assumption that agricultural activities are static with respect to other factors, especially, market driven forces.

have some conception of international relations, the relative significance of other issues that may emerge at some future time, the development of scientific and technological capacities or, the norms and values that will guide social action may bring about incremental changes in temperature which could collectively cause damages or benefits beyond our imagination.[14]

Even if we had a perfect forecast of any gradual changes in surface temperature; even if it were calibrated to socially relevant locations and time frames, we would be unable, given the absence of a theory of society, to analyze the impact if any of such incremental changes on social activities.

An additional deficiency of the models linking social milieus and physical climate changes concerns the unstated assumption of dumb corporate and individual actors; that is, not only will societies look very different from now in a few decades quite independent of climate changes, but individuals and collectivities will respond in a variety of ways to changes in climatic conditions. Such responses can range from trivial adaptations to profound transformations of existential conditions.

Once one is aware of the extensive range of sensible connections (as well as lack of linkages) among social and natural phenomena and of connections among connections, it is insufficient to generally conceptualise and interpret the association between climate and society or climatic changes and social impacts as matter of "stimulus-response". Such an analysis is basically inadequate because it is reductionist; it overemphasises the "rationality" of the response to climatic extremes; for example, the extent to which such responses are driven by exclusively economic rationales excessively simplifies the motives of people. At the same time, it underplays the significance of "irrational" and often ambivalent motives or complexity of responses (cf. Mackay 1981). The models depicting the interaction between social action and natural processes typically do not convey the impression that large-scale social systems are systems in which judgments, emotions, and perception play a salient role. Nor do these models display a concern with changing system properties; for example, trends toward volatility and ungovernability that may, in the end, alter the ways in social systems are capable or incapable of responding to external signals.

This is not to say that the climatologist or the politician might not attempt to eliminate irrational responses to climate changes and devise elaborate campaigns and regulations to reduce reactions to "objective" elements.

---

[14] It is not entirely unreasonable to suggest that individuals and collectivities may actually begin to prefer rather than fear a rise in surface temperature in their parts of the worlds. Most importantly, the dynamics of such preferences will not strictly depend on forecasts about changing surface temperatures. In the same manner, Kellogg (1978:220) cautiously concludes that "a warmer Earth would by no means be a less desirable place to live. He specifies his conclusion with respect to precipitation types and suggests that these patterns "seem to be more favorable to more people than the present ones" but adds "there are some disturbing exceptions such as the Prairie Peninsula in North America." None the less, on the whole, Kellogg (1978:220, 223) concludes, "the prospect for a warmer Earth is a favorable one, though a few places may suffer," but "on the whole the Earth will have a climate more favorable for feeding the increasing population."

However, a more adequate conception of the associations between climate change and society must put such irrational elements squarely in the center of the analysis.

Similarly, researchers cannot simply assume, as some appear to do, that "coping with climate change will occupy much of society's attention over the coming decades" (Malone and Yohe 1992:109). As a matter of fact, attention in the media and among policy makers to the greenhouse effect will fluctuate considerably, as is already the case; and in all likelihood, the focus will vary with climatic extremes (Mazur and Lee 1993).[15]

The models constructed so far do not rely on any historical benchmarks in order to assess the ability of populations to adapt and cope with climate change; for example, as the result of voluntary or involuntary migration. After all, the anticipated increase in mean global temperature is small in comparison to the sometimes dramatic changes in climate that large segments of the world's population experienced in the course of migrating in the western hemisphere in the last centuries (cf. Schelling 1983: 455-456). Even if individuals and groups did not move during their lives, the urbanisation process produced changes in micro-climates that approach or exceed the new anthropogenic changes in climate. More recently, massive waves of migration in the United States; for example, from the northeast and northcentral regions to the south and the west include changes in climate for the migrants that are substantial and would appear to exceed the anticipated changes in average global temperature attributed to increases emissions of carbon dioxide.

Whether the development of historical benchmarks is sensible at all may of course be contested. The possibility of developing such benchmarks may be contested on at least two grounds: (1) It is virtually impossible to factor our relevant dimensions. This may, for example, be due to a lack of data, while data that are available may only refer to periods in which the emancipation by society from climatic conditions was quite significant. The complexity of the relationship between climate and society may, on the other hand, prevent the construction of useful indicators in the first place. (2) The ability to utilise such benchmarks even if they could be developed may be of limited value because the existential conditions have changed in ways that do not allow one to superimpose historical experiences on contemporary or future society (cf. Wigley et al. 1985).

A very common notion found in the discussion of the policy implications of climate models is the conviction that action is needed now and cannot wait until the point where the isomorphism between model and reality has been "perfected" (e.g. Kellogg 1978:209). The urgency of the problem, it is argued, exempts action from the need to conform to "classic" notions of rationality; for example, robust scientific knowledge claims or precise and transparent cost-benefit analyses.

---

[15] The height of the media attention to the issue of global warming occurred during the summer of 1988 in the midst of the worst US drought in fifty years (cf. Mazur and Lee (695)). Only a few years earlier, climatologists (e.g. Kellogg 1978:205) lamented with justification that it is "curious that there is not a more widespread awareness and concern over the potential for altering the planet's environment as a whole," because the media in fact did not pay much attention to the topic.

Virtually all model builders agree that what they call the "complexity" of the task of gauging the impact of climatic variations on social action is enormous. That is to say, the measure of the complexity of complex, and, by inference, of what would constitute the most adequate model, is then a reference to "the effects of climatic variations on all aspects of life in a given society" (Wigley et al. 1985:536). Thus, "*less ambitious investigations*, focusing on one or a small number of economic and social variables, *are more realistic*" (Wigley et al. 1985:536; emphasis added). More realistic means not that the model has a higher degree of isomorphism but is less difficult to generate, at the present.

The existing asymmetries in the construction of the models have to be reduced. For example, the frameworks that are employed cannot move through time in a one-sided manner; that is, the physical signal clearly is moving through time and changing in time while our conception of the social milieu is hardly seen to move through time and change with time. In short, the scientific and political debate on global environmental change stands to benefit from incorporating to a greater extent the insights of the social sciences.

Chapter 2
Models between Academia and Applications

by Hans von Storch

Abstract

In environmental sciences, models are an indispensable tool. However, the seemingly simple technical term "model" covers a wide range of different conceptualisations and images of the real world, ranging from drastic reductions and simplifications to maximum complexity. These different types of models serve different purposes. The reduced, or *cognitive* models constitute "knowledge" while failing to provide detailed descriptions. The other extreme, *quasi-realistic* models create the possibility of simulation and experimentation of real world systems but fail to produce insight into the system's functioning. While fundamental research commonly tends more to cognitive models and applied research to quasi-realistic models, a comprehensive strategy employs both types of models in an interative, synergistic manner.

2.1 Introduction

To start our discourse about models in environmental sciences, we present three cases.

- a laboratory model of sediment resuspension and erosion (Section 2.1.1.)
- a hydraulic model of tides in a semi-closed basin (Section 2.1.2)
- a numerical model of tides in a semi-closed basin (Section 2.1.3)

In the two first cases, the "model" is a mechanical analog of a real situation, whereas the third case is a prototypical purely mathematical "model." After having discussed these examples, we will address some specific aspects of environmental modeling, which makes environmental science different from classical natural sciences (Section 2.1.4). In Section 2.2 general aspects of models are discussed, and the different purposes of *quasi-realistic* and *cognitive* models are considered in Section 2.3. Concluding remarks are given in Section 2.4.

2.1.1   Laboratory Model

The first case considers the morphology of the sea bed and, specifically, its stability in the presence of bottom shear stress generated by waves and/or currents. In most cases this shear stress is generated by turbulent water motion. When there is no turbulence, the surface will remain at rest, whereas heavy turbulence will cause sediment particles to become eroded and resuspended. The details will depend on specifics of the sediment, i.e., like colonisation by diatoms or benthic animals.

For describing the dependency between turbulence and sediment erosion, a simple laboratory set-up has been designed (Schünemann and Kühl 1993; see graphical sketch in Fig. 2.1). A sample of sediment typical for the area of interest is derived and placed in the bottom of a transparent tube; the tube is filled up with water, and a propeller is placed over the sediment sample. The rotation velocity can be set externally; it determines the degree of turbulence. For a better display, the scene is illuminated by a light placed behind the tube. For three rotation frequencies, the effect on erosion and resuspension is shown in Fig. 2.1. For frequencies below a threshold (middle panel in Fig. 2.1), the water column is transparent; after having passed a threshold single particles float in the water (top right panel); and at the highest employed frequency the lower part of the water column has become opaque (bottom right panel). At this time, not only the top layer of the sediment has gone into resuspension, but also the deeper, consolidated sediment is beginning to become mobilised. When the propeller is turned off, the suspended particles slowly deposit again.

This model does not *explain* why the threshold is as it emerges, or how deep the eroded layer is. It "only" informs us about the existence of a threshold and it allows us to determine these critical numbers. To constitute "understanding," we need to abstract from the concrete set-up, and design a conceptual model, a "Gedankenmodell" (mental model). The laboratory model then serves two functions; first to validate the conceptual model, and, second to specify a series of unknown *parameters*. The *conceptual model*[1] pictures turbulence as the key process; the stress associated with the turbulence causes the adhesion at the surface, reinforced by the presence of diatoms, to collapse. When the top layer has disappeared and the turbulence is strong enough then even the consolidated sediment is disintegrated.

2.1.2   Miniaturisation

The second example features a "model" somehow similar to a "toy model" - namely a miniaturised composition which replicates some features of the original. In the case of a child's toy train, the model moves on tracks and wagons are coupled together and drawn by a locomotive and the like. Other real world features

---

[1] Note that there may be several, different, or even conflicting, conceptual models consistent with the laboratory model. In the specific case, one could argue that it is not the turbulence but a vertical drag exerted by the propeller. This hypothesis would be consistent with the specific experiment but is falsified after a closer inspection.

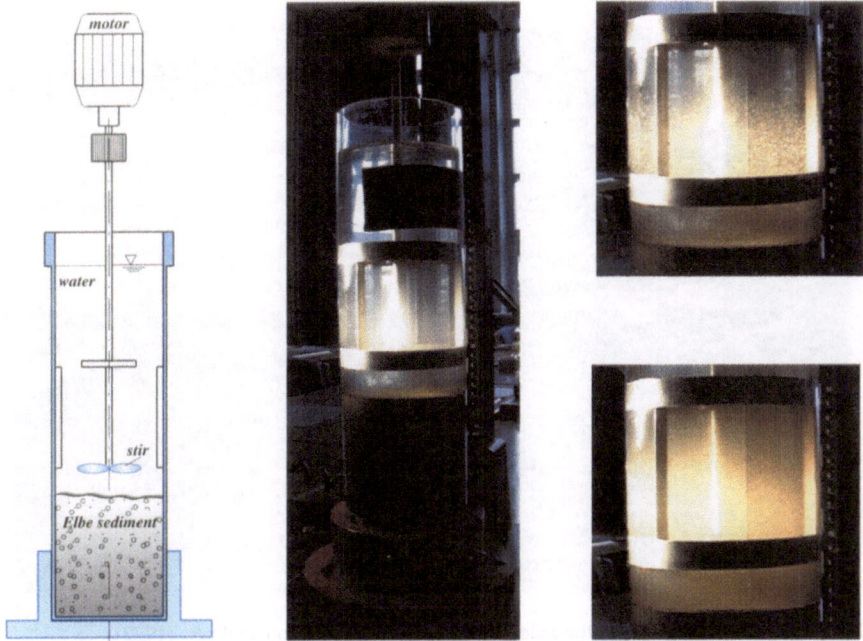

Fig. 2.1. Erosion Stress Laboratory Model. On the left is a sketch of the apparatus, with the sediment in the lower part and 30 cm water on top of it. The propeller generating the turbulence is placed 5 cm above the sediment. In the middle, a photo of the apparatus is displayed, with a very low rotation of the propeller resulting in negligible erosion. In right panels only the water column right above the sediment is shown. In the top panel erosion is about to begin, whereas in the bottom panel the turbulence is so strong so that heavy erosion has been induced.

are not available in the model; for example, in the instance of the toy train, the engine is not really driven by steam but by electricity. Other miniaturised models refer to down-scaled complexes of buildings. In engineering sciences, such miniaturised models have been used extensively in the past. For instance, hydraulic engineering employed huge miniaturisations to replicate the interplay of currents, waves or tides with man-made modifications of rivers or the coast.

Figure 2.2 is a photograph of a downsized model of the Jade Bay in Northern Germany (Sündermann and Vollmers 1972). Is has been scaled to correspond to a real bay with a diameter of about 10 km and with a channel open to the North Sea about 4 km wide. At the open boundary (at the front of the photograph), a sinusoidal tide is imposed. In the basin, currents are displayed by floating white bodies, whose movements appear as white lines on a photograph taken with sufficiently long exposure time. One of such snapshots is displayed in Fig. 2.3; the situation refers to a declining tide with out-flowing waters. The emerging counter- clockwise eddy is marked by two white arrows.

The hydraulic model may be used to provide estimates of the current patterns (Fig. 2.3) as well as of the current speeds during a tidal cycle (Fig. 2.4). The current pattern is symmetric with two eddies in the bay, just before the flow narrows

Fig. 2.2. A hydraulic model of the Jade Bay. From Sündermann and Vollmers (1972)

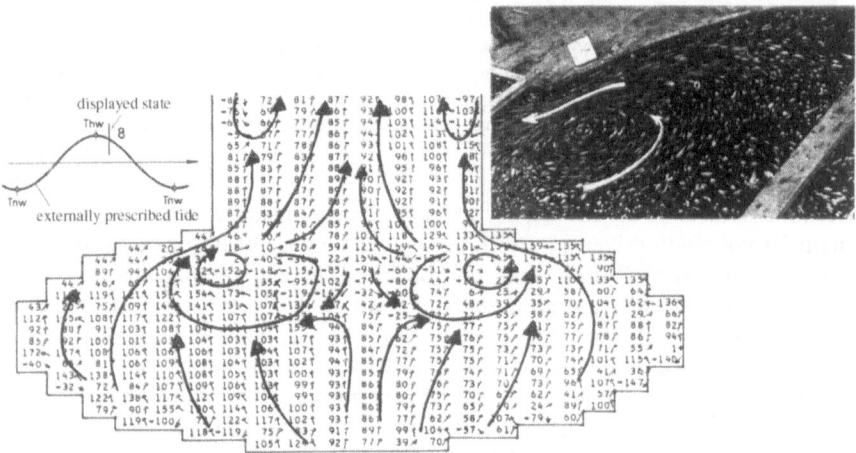

Fig. 2.3. Tidal currents in the hydraulic model (photograph; Section 1-2) and in the numerical model (graph; Section 1-3). The timing is given by the little inset: the tide has just passed the peak level and the water begins to outflow from the basin. From Sündermann and Vollmers (1972)

before entering the channel to the open sea. The speed displays a bi-modal cycle, which is almost symmetric for the inflow- and outflow phases, with maximum speeds of about 70 cm/s.

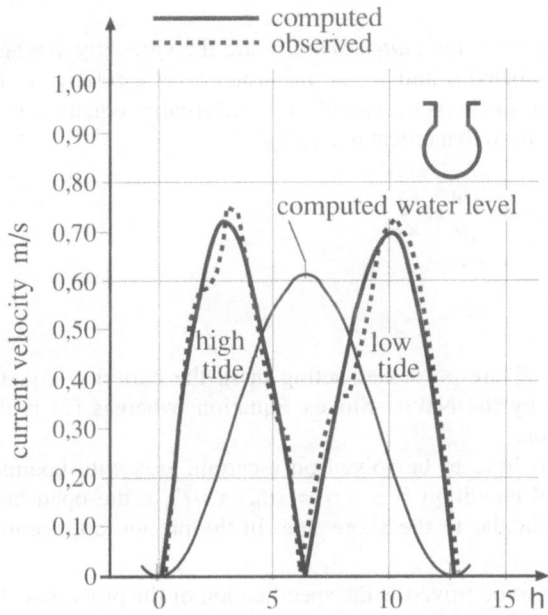

Fig. 2.4. Current speeds and water levels during a tidal cycle in the simulated Jade Bay. The dashed line refers to currents derived form the laboratory experiment (Section 2.1.2), and the solid lines to currents and water levels calculated in the numerical model (Section 2.1.3). From Sündermann and Vollmers (1972)

As in the previous case, the simulation with the hydraulic model does not offer any immediate insight into the *dynamics* of the geophysical system of Jade Bay. It does not tell us *why* there are two opposite eddies. Understanding requires the utilisation of concepts; in this case the principle of conservation of angular momentum. Other questions which remain unanswered are the sensitivity to the depth or the size of the bay. However, the simulation is valuable for the coastal engineer to assess where currents of various magnitudes will appear. Combined with the erosion laboratory model presented in Section 2.1.1, the tidal dependency of suspended matter concentration may, at least in principle be estimated. The value, and purpose of the hydraulic model is to provide a quasi-realistic composition within which certain experiments on the system's sensitivity may be conducted. These simulations provide data which may be used to develop conceptual models and theories about the functioning of the system.

2.1.3 Numerical Models

The same problem of tidal currents in a semi-closed bight, as dealt with in the hydraulic model in the previous Section 2.1.2, has also been dealt with a numerical model. A "numerical" model is a computer code, based on certain mathematical equations (or expressions) after some manipulations such as discretisations or simplifications.

In the present case, the *state variables* are the vertically averaged components of the currents labeled u and v, and the water level ξ relative to the undisturbed level h. For each of the state variables, a differential equation for the change in time is available from dynamical reasoning:

$$\frac{du}{dt} = \sum_i P_i^u \quad \text{and} \quad \frac{dv}{dt} = \sum_i P_i^v \tag{2.1}$$

$$\frac{d\xi}{dt} = \sum_i P_i^\xi \tag{2.2}$$

where P_i^u and P_i^v are processes acting upon the currents. Equations (1) are in principle given by the Navier-Stokes Equation, whereas (2) is the principle of mass conservation.

The equations have to be solved on a certain area with boundary conditions, such as the tidal condition $\xi(x{=}0,t) = sin(2\pi\, t/T)$ at the open boundary and no currents perpendicular to the shore line. In the present application the period is $T = 12.5$ hours.

The next task to be solved is the specification of the processes. The formulation of the net divergence is simply $P_1^\xi = \partial(h+\xi)u/\partial z + \partial(h+\xi)u/\partial y$. The scale analysis of the equations of motion (Pedlosky, 1987) informs that the most important "zero order" processes are the pressure gradient force $P_1^u = g\partial\xi/\partial z$ and $P_2^v = g\partial\xi/\partial y$ and the Coriolis force $P_2^u = -fv$ and $P_2^v = fu$ with the Coriolis parameter f. Another important process is the bottom friction in a turbulent boundary layer. The effect of this process on the state variables "vertically averaged current" can not be described explicitly; instead the effect has to be "parameterised". That means, the average net effect on the current is specified, conditional upon the state of the system in terms of u, v and ξ. A parameterisation is an educated guess, and it is usually adopted after its impact on the overall simulation has emerged as an improvement (for a more detailed discussion refer von Storch, 1999, see also Section 11.3.1). Thus, there may be several rather different formulations for the same process. For bottom friction in shallow waters, the following formulation is often adopted:

$$P_3^u = \frac{ru}{h+\xi}\sqrt{u^2+v^2} \quad \text{and} \quad P_3^v = \frac{rv}{h+\xi}\sqrt{u^2+v^2} \tag{2.3}$$

with a constant friction parameter r. Another process is that of horizontal diffusion, which is often parameterised as

$$P_4^u = A_H\,\Delta u \quad \text{and} \quad P_4^v = A_H\,\Delta v \tag{2.4}$$

with a diffusion parameter A_H and the Laplace operator Δ.

With these specifications, the system is closed; all other processes, be it the effect of variable wind, effects of vertical stratification, the mixing due to shipping, the effect of suspended matter, or the propagation of sound waves in the water are disregarded and considered irrelevant for the problem of tidal currents and water levels. Because of the disregard to all these processes, the model given by equations (2.1-2.4) represents a severely simplified and idealised description of the real world. Using the terminology introduced later in this discourse, the model is an example of a quasi-realistic model, as it has been set up for approximating real tidal currents in spatial and temporal detail. We call it a "mathematical" model.

Before the mathematical model can be implemented on a computer, it has to be discretised. It is transformed from being a infinite dimensional system to a finite system. This is achieved by either replacing the derivatives with finite differences, such as $d\xi/dt \approx \xi(t+\partial) - \xi(t-\partial)/2\delta$, or by expanding the stated variables into a truncated series of orthogonal functions such as the trigonometrics[2] - then the spatial coordinate x is replaced by an index k enumerating the orthogonal functions. Another option is the use of finite elements. In case of the numerical model of Jade Bay, a finite differencing has been adopted. Note that this manipulation further simplifies the model, which we name a "numerical" model.

For what purpose can we use the numerical model? Possible applications are

- attempts to replicate the outcome of the hydraulic model. If both models return similar assessments, they may serve as arguments for the validity of both; if they return conflicting assessments, further analysis is required to decide if one or perhaps both models are "wrong" - in the sense that one or both contradict observational evidence.
- the performance of sensitivity experiments - as for instance: what is the importance of the Coriolis force on the simulated flow regime in Jade Bay? The relative importance of processes, on the formulation of parameterisations and of boundary conditions can be tested.

Both applications have been run with the numerical model of Jade Bay (Sündermann and Vollmers 1972).

In an attempt to validate the hydraulic model, the numerical model was run without invoking the Coriolis force (which could not be considered in the hydraulic model without placing the apparatus on a rotating disk). In Fig. 2.4 the simulated current speed distribution shortly after high tide is shown on the matrix of grid-points; the arrows indicating the directions of the flow are added by hand. The photograph of the laboratory flow with the white lines is consistent with the numerical model. Also the current speeds displayed in Fig. 2.4 are very similar in the numerical and hydraulic model. Based on this evidence, Sündermann and Vollmers concluded that the two approaches return consistent results.

---

[2] If $\xi(x,t) \approx \sum_{k=1}^{K} a_k(t) e^{ikx}$ is the truncated expansion, then the spatial derivative is approximated by $\dfrac{d\xi}{dx} \approx \sum_{k=1}^{K} ika_k(t) e^{ikx}$.

To test the sensitivity of the system to the presence of the Coriolis force, this process was turned on in a second simulation. The resulting current pattern is displayed in Fig. 2.5. Obviously, the current system deviates significantly from the one shown in Fig. 2.3 without the Coriolis force. The currents are no longer symmetric; instead in the left two thirds of the channel the flow is outward, but in one third it is inward. Also, the clockwise eddy on the left has been diminished. Thus, the Coriolis force is found to be a process which should be taken care of; indeed the results obtained with the hydraulic model should be considered with reservations.

The advantage of numerical models over mechanical models is twofold. First, these models are economically much more efficient; the cost of setting up a tank as shown in Fig. 2.2 is by magnitudes larger than setting up a numerical model on a computer. The other advantage is the simplicity to do "observations" in a numerical model; it amounts to adding simple write-commands in the code. These "observations" are accurate, and can be done in high temporal and spatial resolution. Because of this possibility, it was possible to add in Fig. 2.4 water levels simulated in the numerical model, which were "unobservable" in the hydraulic model. However, the ability to get these numbers easily does not mean that the numbers are "right" or meaningful. Instead the numbers *can* be mostly unrelated to the real process, which is supposedly modeled; and *may* reflect to some extent artifacts of the model design. In the present case; however, this seems not to be a problem.

Because of these two advantages, we will consider only mathematical models and their numerical realisations in the following.

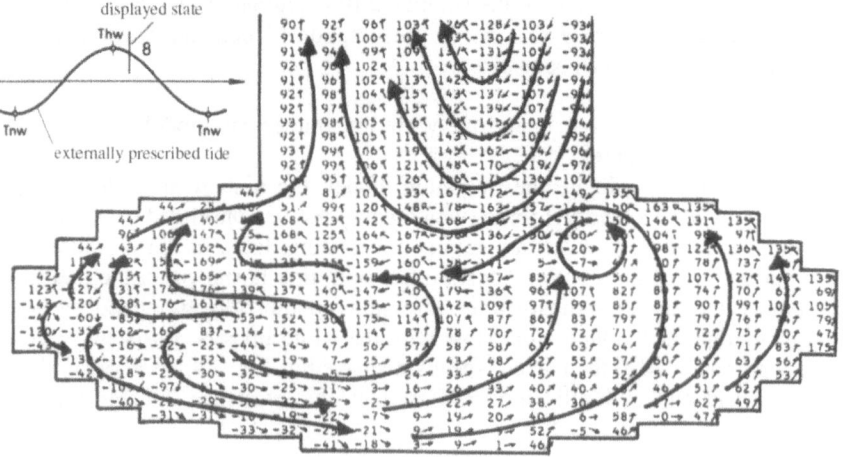

Fig. 2.5. Tidal currents in the numerical model. The timing is given by the little inset: the tide has just passed the peak level and the water begins to outflow from the basin. To be compared with Fig. 2.3. From Sündermann and Vollmers (1972)

2.1.4 Specifics of Environmental Research

Physicists, chemists etc. consider the understanding and prediction of environmental systems just as another physical, chemical etc. problem. Also, everybody

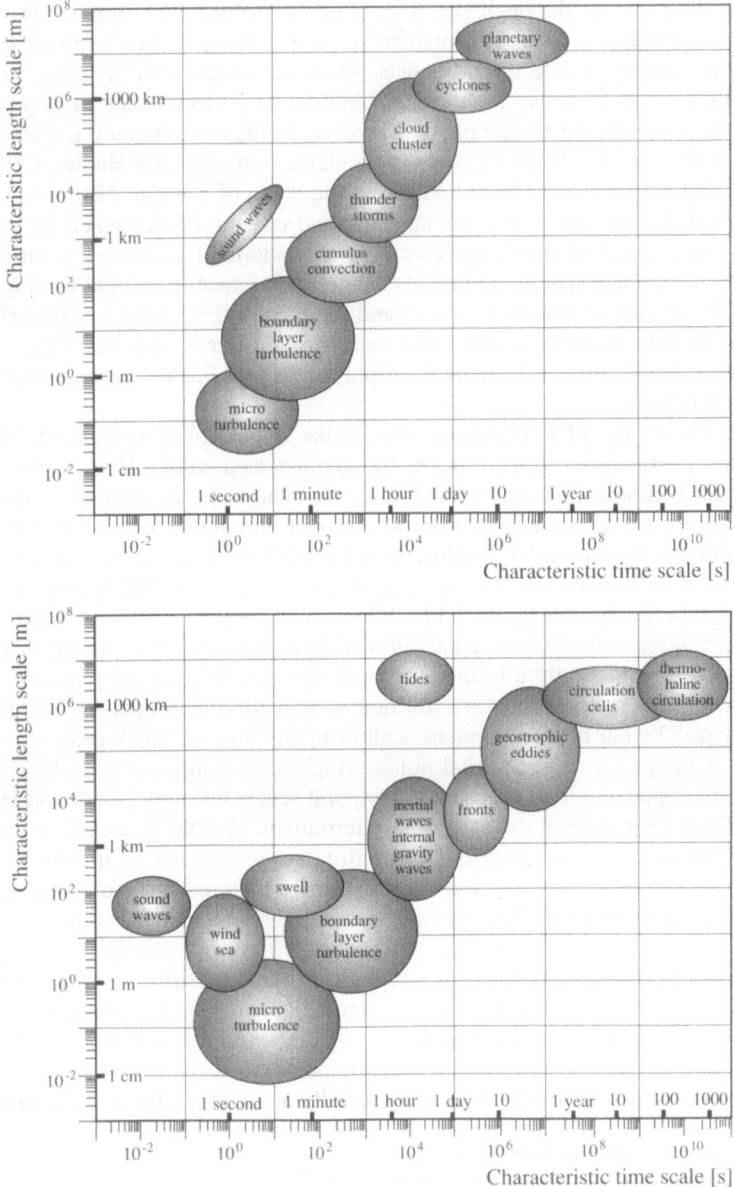

Fig. 2.6. Spatial and temporal scales of atmospheric and oceanic dynamics. From von Storch and Zwiers (1999)

has an intuitive understanding of "the" environment. However, the understanding of the dynamics of environmental systems such as the atmosphere, the ocean, a catchment, or the biosphere; and their interaction requires an approach different than that of a non-scientific lay person or that of the "pure" sciences of physics and chemistry. The scientific employment of the environment poses a number of specific problems (cf. Navarra 1995).

This key difference is the open character of all environmental systems (Oreskes et al. 1994). A myriad of processes interacts in such systems, and they are exposed to an infinite number of external influences. One could argue that the same situation would prevail in a gas, with enormous numbers of molecules interacting with each other and responding to radiations. However, in the environment, the temporal and spatial scales of the processes vary widely, from e.g., the Hadley Cell in the tropical atmosphere to turbulent eddies in the wake of a plane. Moreover, the dynamics at different scales vary in character, and can not be described by some (simple or even complex) similarity laws. Also the external forces are too variable to allow for a complete specification; they range for instance from tidal forcing by the moon, the mixing of waters by a ship and the breathing of people or the effect of diatoms on stabilising the Wadden Sea bottom against erosion stress. This wide range of spatial and temporal scales is displayed in Fig. 2.6 for oceanic and atmospheric dynamics.

There are a number of implications. One is the impossibility to conduct laboratory experiments on the functioning of the *systems* as a whole. Here, following Encyclopedia Brittanica, we understand an experiment as a "an operation carried out under controlled conditions in order to discover an unknown effect or law, to test or establish a hypothesis, or to illustrate a known law." Of course, experiments may be done with sensors on reduced systems[3], but not on the full system. Also, real world *repetitions* are unavailable, which may help to rigorously sort out whether certain phenomena have emerged merely by chance or as a result of certain processes. There is only a limited segment of a trajectory in the phase space; even if the system is ergodic, there are doubts that the phase space is sampled sufficiently well by our limited segment to allow us finding real "analogs."

Second is the presence of internal noise, which is self-organising in the sense that variability appears on all spatial and temporal scales (cf. von Storch and Hasselmann 1996). In principle, the system is deterministic, but the presence of many chaotic processes creates a pattern of variability, which can not be distinguished from random variations.[4] Because of these specific features, two fundamentally different types of mathematical models are used in environmental research:

- One sort is "quasi-realistic" and is supposed to be a substitute reality, within which otherwise impossible experiments can be conducted. A representative of this type is Sündermann's and Vollmers' case discussed above in Section 2.1.4. Such models are also used to extra- and interpolate in a dynamically consistent manner the sparse observations, so that spatially and temporally high resolution

---

[3] As with the stability of the sediment discussed in Section 2.1.1.

[4] In the case of the tides in Jade Bay, this aspect was not relevant; as only a periodic, purely deterministic forcing was applied, and the considered system is not chaotic but strongly dissipative because of the bottom friction and horizontal diffusion.

analyses of the system's state are constructed (in particular weather analyses, e.g. Kalnay et al. 1996).

- The other type of model, named here "cognitive", is highly simplified and idealised. Because of its reduced complexity, such a model constitutes "knowledge". The geostrophic model $P_1^u = -P_2^u$ and $P_1^v = -P_2^v$ is an example of this type of model.[5] Other examples are Lorenz' chaotic system (Lorenz 1963) or Hasselmann's stochastic climate model (Hasselmann 1976).

In the following we will discuss these two types of mathematical models in some more detail.

2.2 General Properties of Models

Models are supposed to reflect reality. As such they deviate from reality, as the introductory examples have demonstrated.

Models are *smaller, simpler* and closed in contrast to reality, which is always *open*. This difference is attempted to be sketched in Fig. 2.7 and 2.8.

- "Smaller" means that only a limited number of the infinite number of real processes can be accounted for. In the case of the tidal model, processes related to varying density were disregarded; also topographic details on spatial scales smaller than the grid cells' size could not be described. In fact, only the processes P_1 - P_4 were considered.

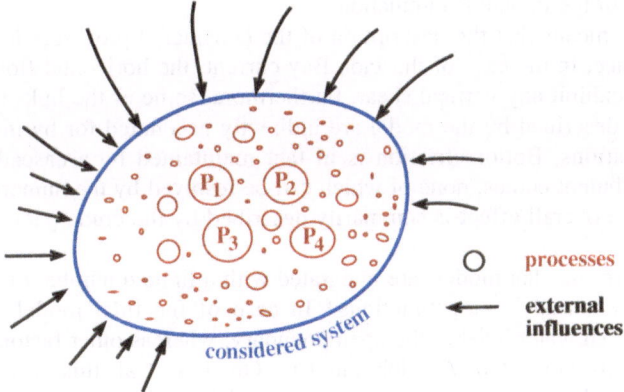

Fig. 2.7. Sketch of a real system, in which an infinite number of processes P_i (open circles) is present, and upon which an infinite number of external forces (arrows) act

---

[5] For the notation, see Section 2.1.3.

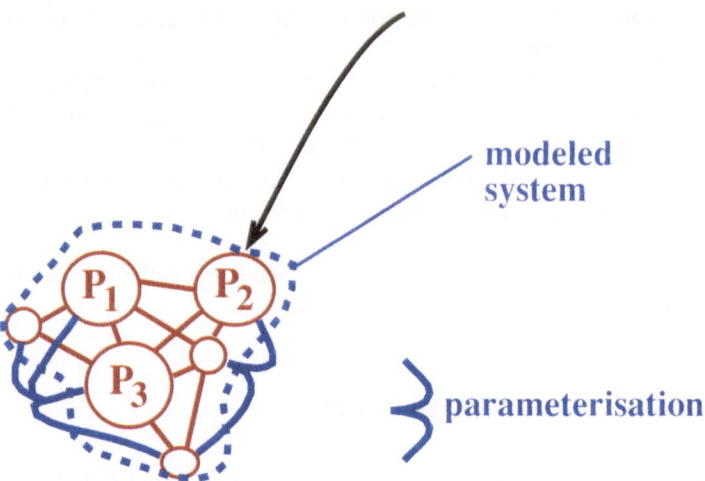

Fig. 2.8. Sketch of a modeled system, in which only a limited number of processes (open circles) and their interactions are represented, and in which the number of external forces is also limited (arrow). Parameterisations are indicated by solid lines crossing the dashed-line border of the model

- In the case of an atmospheric model or an oceanic model, the unavoidable discretisation means that from the overall ranges of scales, as displayed in Fig. 2.6 only a limited interval can be accounted for. A global model describes planetary waves and cyclones, but no boundary layer turbulence in any detail. Similarly, an ocean model resolving internal gravity waves will hardly describe the dynamics of thermohaline circulation.
- "Simpler" means that the description of the considered processes is simplified. For instance, in the case of the Jade Bay current, the horizontal flow is not allowed to exhibit any vertical shear. Furthermore, some of the links to the processes not described by the model are indirectly accounted for by means of parameterisations. Bottom friction is in fact maintained by a cascade of small scaled turbulent eddies, none of which can be resolved by the numerical model. Instead the overall effect is summarily described by the crude parameterisation (2.3).
- "Closed" means that models are integrated with a limited number of completely specified external forcing functions:[6] In case of the tidal model, it was the tidally driven water level at the open boundary, whereas other factors like wind forcing were neglected. As elaborated by Oreskes et al. this is an important philosophical limitation of environmental models, as it implies that the "right" answer of a model may be due to either the "correctness" of the model or an coincidental balance of an incorrect model response and the effect of an unaccounted external influence.

---

[6] An exception represent models in which randomised external influence factors are specified. An example is provided by Mikolajewicz and Maier-Reimer (1990)

Because of these properties of models, they suffer from a number of limitations:

- A model desribes only part of reality. For instance, the numerical tide model described above is limited to time scales of a tide, to an area small enough to allow for the assumption of a constant Coriolis coefficient and to water bodies of a minimum depth. In its present form it can not be used for predicting water level variations due to meteorological variations or due to runoff from rivers.
- This limitation is sketched in Fig. 2.9 in a space-time-parameter phase space. When setting up a model, the researcher almost always makes assumptions about the time and space scales, and about the range of parameters. In the best case, these assumptions are made explicit, but often they are implicit and unaware to users of models and model outputs.
- Indeed, the choice of the "admissible domain" is a subjective process; ideally it is guided by a rigorous analysis of the relative importance of different processes on different scales. A classical approach to this end is to transform the equations first into a dimensionless form, featuring time and space scales as well as characteristic parameters explicitly. Then, a Taylor expansion allows to discriminate the various terms according to their relative importance (cf., Pedlosky 1987).
- The models can not be verified in the sense that we can with certainty conclude that the model is producing "right" numbers because of the "right" reasons (dynamics).[7] We can compare the numbers with observed numbers and conclude that they are consistent with the observations[8]; we may add to the credibility of the model by analysing the dynamical system and assuring that all first order processes are adequately accounted for. In that case we call the model *validated*. In that sense we may trust the model's output as long as we are applying it within the "admissible" domain depicted in Fig. 2.9. We may be confident that the model may be used for some extrapolations (in Fig. 2.9, an application to the point A would amount to an extrapolation), such as the effect of dredging the inflow channel, but we can not derive knowledge about the system's response to, e.g., making the inflow channel very shallow.
- The problem of making statements *outside* the admissible domain is frequently met in applications. The various claims about anthropogenic changes of climate are based on such extrapolations of models. For instance, if a model is realistic in reproducing the present climate, it is not assured that its response to changing greenhouse gas concentrations is described realistically.

---

[7] We will not discuss the meaning of "right" in this context.

[8] "Comparing with observations" is a trivial act if it is reduced to compare an observed map or curve with a simulated map or curve. In general, however, this naive approach is insufficient. When the forecasting capability is indicative of a model's skill, then ensembles of forecast should be formed and overall measures of success be calculated (e.g. Livezey 1995); when the system is unpredictable, then a statistical comparison of simulated and observed data is required; for instance in terms of means, characteristic patterns and spectra (von Storch and Zwiers 1999).

2.3 Purpose of Models

2.3.1 Quasi-realistic Models Surrogate Reality

The main purpose of a quasi-realistic model is to provide scientists with an experimental tool. As such, it is as complex as possible. A quasi-realistic model generates numbers as detailed as the real world (within the limits of spatial and temporal resolution). These numbers are consistent with observed numbers, i.e., to some approximation they are (statistically) indistinguishable from observed numbers. Prototypes are modern climate models, featuring detailed dynamical models of the atmospheric, oceanic and (part of) the cryospheric dynamics. But even if highly complex, it is limited to its specific "admissible" domain in terms of scales and parameters (Fig. 2.9).

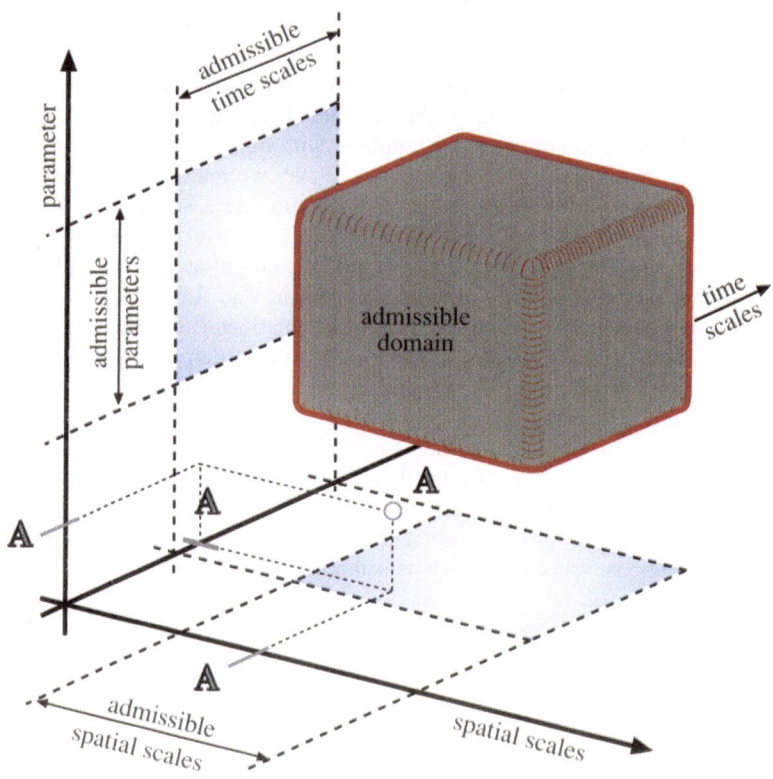

Fig. 2.9. Admissible domain of applicability of a model. Application to point A would be an extrapolation.

Quasi-realistic models are almost always expressed in numerical terms, and rarely as mechanical or other analogs. Essentially, they represent an engineering approach.

A quasi-realistic model is composed of many sub-models, describing the various processes involved in the dynamics of the considered system. These sub-models, or process-models are ideally quasi-realistic models. In many cases, however, the sub-models are strongly reduced to simple approximations and parameterisations.

Quasi-realistic models are used for various purposes:

- They are used to test hypotheses, for instance to determine which of two concurrent processes is more relevant in certain situations, and to quantify the sensitivity of the system to parameters, like the water depth in the case of the tidal basin or the greenhouse gas concentration in the climate change problem.
- Another application is simulation. This accounts for the derivation of scenarios, i.e., of possible future developments given certain changes in the system's ambient conditions. Such scenarios play a crucial role in many managerial decision-making processes, ranging from coastal engineering to climate policy.
- A somewhat different application is the performance of "control runs", i.e. of running the model with real or statistically modeled boundary conditions with the purpose of generating long time series of complete and dynamically consistent data. In particular, climate models are used in this manner, since detailed observations of the deep ocean or the free atmosphere are scarce or only have been available for a few decades. Then, the output of such control runs serves as substitute observations and is used to derive hypotheses about the real world's functioning. In certain applications, such data are also used in managerial decisions, as for instance concerning risk assessment in safety design for off-shore constructions.
- Forecasting of the near-future development of the system is also done with such models. Numerical Weather Prediction models belong into this category.
- A relatively modern application concerns the analysis of environmental states. Because of many degrees of freedom and practical barriers rendering certain variables unobservable, a complete analysis of the state's system is all but impossible. However, an intelligent use of dynamical knowledge encoded in quasi-realistic models allows for the dynamically consistent interpretation of sparse, and to some extent uncertain, observations (Robinson et al. 1998). Such tools are called *data assimilations*, and their output "analyses". Note that analyses are merely best guesses of the real situation; it is a skillful approximation of the real situation; sometimes measures of certainty of the approximations are given.

In this way, consistent and complete data bases of the system are provided.

Quasi-realistic models do not provide immediate knowledge. Being most complex, the numbers need a skillful analytical treatment before conclusions can be drawn. The knowledge is hidden in the numbers; to extract this knowledge the output must be interpreted with the help of cognitive models, i.e., concepts derived from dynamical reasoning, screening observations, previous numerical experimentation, or simulation with statistical techniques (von Storch and Zwiers 1999).

2.3.2   Cognitive Models: Reduction of Complex Systems

Cognitive models are characterised to be of minimum complexity; they describe all processes of "first order," i.e., all processes which are required to describe the main features of interest. The description of these processes is stripped down to the bare essentials. As such, cognitive models constitute knowledge. When using the phrase "we understand a system" (for certain scales and parameter ranges), we actually mean "we have a cognitive model to describe the phenomena we observe (or expect)." As such, the formulation of cognitive models is a key method in fundamental science.

A prototypical case is the zero-dimensional energy-balance model of climate (e.g., Crowley and North 1991), in which the Earth's surface temperature is described as being in balance between incoming short wave radiation and outgoing long-wave radiation, and the two processes modulated by albedo and back scattering. Often, the needed parameters in such models are determined semi-empirically.

Of course, cognitive models may also be built with the intention to derive hypotheses, i.e., by suggesting certain processes to be of first order and to derive hypotheses about implications. Then these hypotheses may be examined with the help of observational evidence or simulations with quasi-realistic models.

In most cases, the derivation of cognitive models is left to the insight and ingenuity of the researcher. A classical case are the two rivaling explanations of the Gulf Stream put forward by Munk and Stommel in the late 1940s (see Pedlosky 1987). However, there are ways of pursuing the goal of a "good" cognitive model in an objective manner. The scale analysis mentioned above, based on a Taylor expansion of the relevant parameters in the dimensionless equations is one such way. Another general one is Hasselmann's *Principal Interaction Pattern* concept (Section 11.4: Hasselmann 1988; von Storch and Zwiers 1999).

2.4 Conclusions

In this discourse we have discussed the scientific approach of "modeling," which is usually not conceptualised. In about all scientific disciplines, classical natural sciences, environmental sciences, social and cultural sciences, the term "model" is used. A joint property of all these models is that they refer to a complex part of reality and that they are simpler than reality. Otherwise, these models vary widely in concept, design and purpose. Nevertheless everybody seems to believe that his or her use of the term is the genuine one, supposedly understood by everybody else. Examples are mental maps in social sciences, digital elevation maps in earth science and world models in economy. Some models are static, like a map, others are dynamic, including a predictive capability. Some models are scientific constructs, others are social or historical constructs.

In interdisciplinary cooperation, then, severe misunderstandings emerge and hinder the flow of ideas and knowledge between the different traditional branches of science. This is in particular a problem in modern environmental sciences, which are rapidly expanding across traditional disciplinary borders.

In this paper we have attempted to characterise two major types of models employed in physical environmental sciences; cognitive and quasi-realistic models. Both play a key role in the progress of environmental science. Indeed, in this science the classical loop "... --→ *experiment* --→ *theory* --→ *experiment* --→ *theory* --→ ..." is replaced by the loop "... --→ *quasi-realistic* --→ *cognitive* --→ *quasi-realistic* --→...".

Acknowledgments

I am grateful to Götz Flöser, Herbert Kühl and Walter Puls for providing me with background information about the erosion experiment, and to Beate Gardeike for professionally preparing the diagrams. Fruitful discussions with Peter Müller helped to clarify the overall approach.

Chapter 3
Basic Concepts in Dynamical Modeling

by Peter Müller

Abstract

This paper reviews some of the physical theories that form the basis for the dynamical modeling of environmental systems. The theories apply to different idealized circumstances and employ different physical and mathematical concepts. The review emphasizes the distinction between discrete and continuous, conservative and dissipative, reversible and irreversible, deterministic and stochastic, predictable and chaotic, Lagrangian and Eulerian, and microscopic and macroscopic systems or descriptions. It is shown that the introduction of probabilistic concepts cannot resolve the incompatibility of reversible microscopic dynamics and irreversible macroscopic dynamics.

3.1 Introduction

Numerical models of the physical environment are generally based on a set of dynamical equations that are supposed to represent the laws of physics or "the physics of the system." Laws of physics can be expressed in compact form, e.g. as differential equations, and their validity can be demonstrated in "simple" experiments. The application of these laws to the environment, which is a complex and open system, is by no means trivial and represents a major problem in environmental modeling. The reason is that physical theories offer many ways to formulate dynamical equations, using different concepts and representations. The state of the system can either be specified by the position of a set of discrete particles or by the values of a set of continuous field variables. The dynamics can be Hamiltonian and reversible or dissipative and irreversible. Field theories can be formulated in Lagrangian or Eulerian frames of reference. The equations can be deterministic or stochastic. Systems can be predictable or chaotic. One can choose a microscopic or macroscopic description. These and other basic concepts are reviewed here as they manifest themselves in physical theories such as classical mechanics, thermodynamics, and statistical mechanics. It is the basic thesis of this article that there are no dynamical equations for environmental systems per se. The fact that the system must obey the laws of physics does not automatically yield a set of useful dynamical equations. The modeler must choose a specific set of dynamical equations or a specific physics. This article is to demonstrate that there are indeed many kinds of physics. The choice of physics will depend on prior knowledge of the functioning of the system and on the objectives of the modeling effort. The choice of physics is the first decision a modeler makes. It is a major decision. Different choices will lead to different results.

3.2 Classical Mechanics

Classical mechanics is the oldest of the physical theories. It was put on a quantitative basis by Newton.

3.2.1 Equations of Motion

The standard problem in classical mechanics is the motion of a particle under the action of forces. In classical mechanics, a particle is a point of mass m. Its motion is described by its position x in physical space as a function of time t

$$x = x(t) \tag{3.1}$$

This function is called the particle trajectory. The dynamical law governing the motion of a particle is Newton's second law

$$m\ddot{x}(t) = F(x) \tag{3.2}$$

which states that the mass m times the acceleration $\ddot{x}$ is given by the force F acting on the particle. The dot denotes a time derivative. Often the force can be derived from the potential energy $U(x)$ of the particle by $F = \nabla U$. Equation (3.2) constitutes the equation of motion for the problem. Together with initial conditions for the position and velocity

$$x(t = 0) = x_0 \tag{3.3a}$$
$$\dot{x}(t = 0) = v_0 \tag{3.3b}$$

it can be solved to give the particle trajectory $x(t)$.

The simplest example is that of a one-dimensional harmonic oscillator whose equation of motion is given by

$$m\ddot{x} = -Dx \tag{3.4}$$

where D is the "spring" constant. The solution is

$$x = \frac{v_0}{\omega} \sin\omega t + x_0 \cos\omega t \tag{3.5}$$

with the frequency ω given by

$$\omega^2 = \frac{D}{m} \tag{3.6}$$

Some noteworthy properties of the equation of motion (3.2) are these:

- It is deterministic
- It is reversible. If $x_1(t)$ is a possible solution, so is $x_2(t) = x_1(-t)$.

Newton's law implies a somewhat "static" view of the dynamical evolution of the system. The past and the future of the particle are completely determined by the equation of motion and by the initial conditions. There is no real change. This "static" character of classical mechanics has been a problem in the philosophical discussion of concepts like evolution and free will.

The extension to a N-particle system is straightforward and given by

$$m_n \ddot{x}_n(t) \;=\; -\nabla_n U_i\big(x_1,\ldots,x_N\big) - \nabla_n U_e(x) \qquad n = 1,\ldots,N \tag{3.7}$$

where the index n counts the particles and where, following standard procedures, the potential has been split up into an interaction potential $U_i(x_1,\ldots,x_N)$, which depends on the positions and possibly on the velocities of all interacting particles and into an external potential $U_e(x)$ that is prescribed. Typical applications are the N-body problem in celestial mechanics where $N = O(10)$ and the microscopic description of gases where $N = O(10^{23})$. Another extension is the dynamics of rigid bodies where in addition to the position of the rigid body one needs variables that describe the orientation of the rigid body.

3.2.2 Hamiltonian Dynamics

The equation of motion (3.2) is a second order ordinary differential equation. By the coordinate transformation

$$q = x \tag{3.8a}$$
$$p = m\dot{x} \tag{3.8b}$$

it can be transformed into a coupled set of first order differential equations

$$\dot{q} = \frac{p}{m} \tag{3.9a}$$
$$\dot{p} = F \tag{3.9b}$$

The variable p is called the momentum.

The Hamiltonian H is the energy expressed in these new variables. For the system (3.2), the Hamiltonian is

$$H(q, p) = \frac{1}{2m} p \cdot p + U(q) \tag{3.10}$$

where the first term is the kinetic and the second term is the potential energy. For the one-dimensional harmonic oscillator (3.4), we specifically have

$$H = \frac{1}{2m} p^2 + \frac{1}{2} Dq^2 \tag{3.11}$$

Using the Hamiltonian, the equations of motion (3.9) can be written in the form

$$\dot{q} = \frac{\partial H}{\partial p} \tag{3.12a}$$

$$\dot{p} = -\frac{\partial H}{\partial q} \tag{3.12b}$$

In this form they are called Hamilton's equations.

In the Hamiltonian description of classical mechanics, all the relevant dynamical information is contained in one single scalar function, the Hamiltonian, even for an N-Particle system. This is quite a compact description.

A solution $(q(t), p(t))$ of Hamilton's equations is now a trajectory in the six-dimensional phase space spanned by q and p. For an N-particle system, the phase space has dimension $6N$. For a single particle moving in one direction, the phase space is two-dimensional, and Hamilton's equations are equivalent to the equations

$$\dot{x} = u = \frac{\partial \psi}{\partial y} \tag{3.13a}$$

$$\dot{y} = v = -\frac{\partial \psi}{\partial x} \tag{3.13b}$$

that describe the flow of an incompressible fluid in two dimensions with stream function $\psi(x,y)$.

3.2.3 Integrable Systems

A system is called linear if the equations of motion are linear in the variables q and p. The Hamiltonian is then a quadratic form. The harmonic oscillator (3.11) is a linear system. A quadratic Hamiltonian can generally be diagonalised by a suitable coordinate transformation. In the new variables, the system then represents a set of uncoupled linear oscillators.

Another simple situation arises when a set of "angular" and "action" variables (θ, J) can be found such that the Hamiltonian is a function of the action variables J only:

$$H = H(J) \tag{3.14}$$

Then the system is called integrable. Indeed, in this case, Hamilton's equations take the form

$$\dot{\theta} = \frac{\partial H}{\partial J} = \omega \ (J) \tag{3.15a}$$

$$\dot{J} = -\frac{\partial H}{\partial \theta} = 0 \tag{3.15b}$$

and are easily integrated to yield

$$J(t) \ = \ J_0 \tag{3.16a}$$

$$\theta(t) = \omega(J_0)t + \delta_0 \tag{3.16b}$$

with J_0 and δ_0 being constant vectors. The harmonic oscillator is, of course, such an integrable system.

The integrable system (3.14) has three constants of motion, the three components of the vector J. This constrains the motion to a three-dimensional subspace of the six-dimensional phase space. For the one-dimensional linear oscillator, the motion is constrained to an ellipse (or circle if properly normalised) in two-dimensional (q,p)-space. For two uncoupled linear oscillators, the motion is constrained to the two-dimensional surface of a torus in four-dimensional (q_1,p_1,q_2,p_2)-space. For an N-particle system, the motion is constrained to a $3N$-dimensional subspace.

For most of the last century, physicists have searched for the magic transformation that would turn their equations of motion into the integrable form (3.15). At the end of the century, they finally realised that most interesting systems, starting from the celebrated three-body problem, are not integrable. They generally have only one constant of motion; the energy. The motion is thus only constrained to the $(6N\text{-}1)$-dimensional energy surface in phase space.

Furthermore, most systems are not only not integrable, but they are also ergodic. Any trajectory comes arbitrarily close to any point on the energy surface. The trajectory "fills out" the energy surface. Filling out the $(6N\text{-}1)$ dimensional energy surface (as ergodic systems do) is quite different from being constrained to a $3N$-dimensional surface (as integrable systems are), especially if $N = O(10^{23})$.

3.2.4   Flow in Phase Space

A single trajectory is not enough to characterise a system and to deal with situations where the initial conditions are not known exactly. One must consider the ensemble of trajectories or the time dependent mapping

$$q = q(q_0, p_0, t) \tag{3.17a}$$

$$p = p(q_0, p_0, t) \tag{3.17b}$$

which maps initial values (q_0, p_0) onto their values at some future (or earlier) time. This mapping describes a flow in phase space (or on the energy surface) much like a fluid flow in physical space. For Hamiltonian systems, this flow in phase space has zero divergence

$$div(\dot{q}, \dot{p}) = \frac{\partial^2 H}{\partial q \, \partial p} - \frac{\partial^2 H}{\partial p \, \partial q} = 0 \tag{3.18}$$

The volume in phase space is conserved by the motion. This is Liouville's theorem.

If the system is nonlinear, the geometric and topological properties of the mapping (3.17) or the associated flow in phase space might be quite complex. A given volume element might be stretched out into ever finer filaments until the whole phase space is covered. Neighbouring points diverge and end up in different parts of the phase space. Such flows are called mixing flows. More formally they are defined as follows. Consider a volume element A_0 at the time $t=0$. It is mapped onto the volume element A_t at time t. Denote by P the relative volume of these volume elements, i.e., their volume divided by the total volume of phase space. If the volume element A_0 is asymptotically stretched out over the whole phase space, then one has for any volume element B

$$\lim_{t \longrightarrow \infty} P(A_t \cap B) = P(A_0)P(B) \tag{3.19}$$

The question is, of course, what is being mixed here, since the system is reversible? First, equation (3.19) also holds when we let t approach minus infinity. The time symmetry is not broken. What is being mixed or deteriorating is "coarse grained" information about the system. This can be seen if we interpret P as a probability and apply (3.19) to $B = A_0$. We then find

$$\lim_{t \longrightarrow \infty} \left\langle \left(y(t) - \langle y(t) \rangle \right) \left(y(0) - \langle y(0) \rangle \right) \right\rangle = 0 \tag{3.20}$$

where $y(t) = (q(t), p(t))$. Cornered brackets denote an ensemble average or average over phase space. Equation (3.20) states that the values $y(t)$ become less and less correlated with the initial value $y(0)$ as time progresses.

3.3 Ideal Fluids

Ideal or perfect fluids do not incorporate any dissipative processes. Their theory is an example of a classical field theory. Ideal fluid mechanics is formulated in either a Lagrangian or an Eulerian frame of reference.

3.3.1 Lagrangian Description

In the Lagrangian description, the ideal fluid flow is described by the position x of a fluid particle as a function of its label s and time τ

$$x = x(s,\tau) \tag{3.21}$$

The label coordinates $s = (s_1, s_2, s_3)$ form a three-dimensional manifold. Often the initial condition $x_0 = x(s, \tau=0)$ is used as a label. The Lagrangian description represents a time-dependent mapping from label to position space. The equations of motion for an ideal fluid in an external gravitational potential $\phi_g(x)$ are given by

$$\rho \ddot{x} = -\nabla p - \rho \nabla \phi_g \tag{3.22}$$

where ρ is the mass density and p the pressure. The dot again denotes the time derivative. The mass density is inferred from the conservation of mass

$$\rho(s,\tau) \;=\; \frac{1}{K} \rho(s, \tau = 0) \tag{3.23}$$

where $K = \partial(x)/\partial(x_0)$ is the Jacobian of the mapping from the initial position x_0 to the actual position x. The pressure is given by the equation of state

$$p = p(\rho, \eta) \tag{3.24}$$

where we have assumed a one-component system and chosen the specific entropy η as the second independent thermodynamic variable. For an ideal fluid, the specific entropy of a fluid particle does not change with time

$$\eta(s,\tau) = \eta(s, \tau = 0) \tag{3.25}$$

This is the definition of an ideal fluid. The equation of motion (3.22) together with (3.23), (3.24), and (3.25) provide a complete dynamical description of a perfect fluid.

The Lagrangian description is the generalisation of the discrete N-particle system (discussed in the previous section) to a continuous system with an infinite number of degrees of freedom. The discrete particle label i is replaced by the continuous label s. The equation of motion (3.22) is a continuous version of (3.7). The pressure is the "interaction potential" that describes the effect of all neighbouring fluid particles on the particle under consideration. Most of the concepts of classical mechanics can be generalised to ideal fluids. Ideal fluids are Hamiltonian systems. The Hamiltonian density is given by

$$H = \left[\frac{1}{2} \dot{x} \cdot \dot{x} + e(\rho, \eta) + \phi_g(x) \right] \tag{3.26}$$

where the first term is the kinetic energy density, $\rho\, e\,(\rho,\eta)$ the internal energy density, and $\rho\phi_g$ the potential energy density. However, fluid dynamicists rarely use the Lagrangian description, despite the fact that it represents a straightforward generalisation of classical mechanics. Instead, they use the Eulerian description.

3.3.2 Eulerian Description

In the Eulerian description, the position x and time t are introduced as independent variables and the fluid velocity

$$u(x,t) = \dot{x}(s,\tau) \tag{3.27}$$

as the dependent variable. To avoid ambiguities, time is denoted by t in the Eulerian frame and by τ in the Lagrangian frame. Lagrangian and Eulerian time derivatives are related by

$$\frac{\partial}{\partial\tau} f(s,\tau) = (\frac{\partial}{\partial t} + u\cdot\nabla) f(x,t) =: \frac{D}{Dt} f(x,t) \tag{3.28}$$

where D/Dt is called either the advective, the convective, the Lagrangian, or the material time derivative. Using (3.28), the momentum, mass, and entropy equations (3.22), (3.23), and (3.25) transform to the Eulerian equations

$$\rho\frac{D}{Dt} u = -\nabla p - \rho\nabla\phi_g \tag{3.29a}$$

$$\frac{D}{Dt}\rho = -\rho\nabla\cdot u \tag{3.29b}$$

$$\frac{D}{Dt}\eta = 0 \tag{3.29c}$$

where all variables are functions of x and t. These equations have again to be augmented by the equation of state $p = p(\rho,\eta)$.

The Eulerian equations (3.29) are true field equations. They are a set of partial differential equations with respect to space x and time t. They are deterministic and reversible. They are simpler than the Lagrangian equations. The Eulerian equations can be solved (in principle) without calculating the particle trajectories. If one is interested in the particle trajectories, one can calculate them from the Eulerian velocity field $u(x, t)$ by solving the ordinary differential equations

$$\frac{d}{dt}x(t) = u\big(x(t),t\big) \tag{3.30}$$

subject to the initial conditions $x(t = 0) = x_0$.

The Eulerian description is a reduction of the dynamics. Instead of the six canonical variables x and $p = \rho\,\dot{x}$ in the Lagrangian frame, one has only the five variables u, ρ, and η in the Eulerian frame. The price paid for this reduction is that these variables are non-canonical. The reason why this reduction is possible is that the thermodynamic state of a fluid particle is completely determined by two thermodynamic parameters, say ρ and η, whereas the Lagrangian description uses a three-dimensional manifold $s=(s_1,s_2,s_3)$ to label fluid particles. A one-dimensional relabeling transformation can be carried out without affecting the thermodynamic state of the fluid particle and hence without affecting the dynamics of the system. Any transformation that does not change the dynamics gives rise, via Noether's theorem, to a conservation law. The conservation law associated with the invariance under particle relabeling is the conservation of potential vorticity. This conservation law has no analogue in discrete particle mechanics, since Noether's theorem requires infinitesimal continuous transformations.

The Eulerian field equations can be cast into a Hamiltonian form by adding auxiliary non-physical variables. A Hamiltonian formulation always requires an even number of variables. Because of this addition of non-physical variables Hamiltonian formulations of the Eulerian equations are rarely used.

3.4 Thermodynamics

Macroscopic systems are governed by the laws of thermodynamics. Thermodynamics is quite different from classical particle mechanics. Macroscopic systems are dissipative and irreversible. They are under the spell of the second law of thermodynamics.

3.4.1 The Second Law of Thermodynamics

The second law of thermodynamics considers changes in the entropy S of a system. The total change of entropy is given by

$$dS = dS_e + dS_i \tag{3.31}$$

where

$$dS_e = \frac{\delta Q}{T} \tag{3.32}$$

is the entropy change caused by the exchange of heat δQ with the surroundings. Here T is the absolute temperature. The entropy increases if heat is added and decreases if heat is subtracted. The entropy changes by internal processes are given by dS_i. The second law of thermodynamics states that these internal processes always increase the entropy

$$dS_i \geq 0 \tag{3.33}$$

The processes that cause the entropy increase are the dissipative processes of molecular friction, conduction, and diffusion. Processes for which $dS_i = 0$ are called reversible; processes for which $dS_i > 0$ are called irreversible. In the theory of macroscopic systems, reversible processes are an unattainable limit.

The second law implies that an isolated system will approach an equilibrium state where the entropy is a maximum, subject to the constraints that the system conserves its total mass, momentum, angular momentum, and energy. In thermodynamic equilibrium, the temperature and the chemical potential are constant, and the system as a whole moves with a constant translational velocity and rotates with constant angular frequency.

The second law states that macroscopic systems are irreversible. This is not compatible with reversible classical particle mechanics.

3.4.2 Diffusion

A particularly simple example that demonstrates the existence of an equilibrium state and the approach towards equilibrium is the macroscopic diffusion process. Macroscopic diffusion (in one dimension) is governed by Fick's law

$$\partial_t c = D \partial_x \partial_x c \tag{3.34}$$

where $c\,(x,t)$ is the concentration of a substance and D (= const) the diffusion coefficient. If the diffusion occurs in a "container" of unit length, then the boundary condition that no mass leaves the "container" is

$$D\,\partial_x c = 0 \quad \text{at} \quad x = \pm 1/2 \tag{3.35}$$

First, there exists an entropy functional for this system, namely

$$S = -\int_{-1/2}^{+1/2} dx (\partial_x c)^2 \le 0 \tag{3.36}$$

This entropy changes by

$$\frac{dS}{dt} = -\int dx 2 \partial_x c \partial_x \partial_t c \tag{3.37}$$

$$= -2\int dx \partial_x c \partial_x (D \partial_x \partial_x c)$$

$$= -2D \int dx \partial_x (\partial_x c \partial_x \partial_x c) + 2D \int dx (\partial_x \partial_x c)^2$$

$$= 2D \int dx (\partial_x \partial_x c)^2 \ge 0$$

where we have used the diffusion equation (3.34) and the boundary condition (3.35). Thus the system evolves from some initial state where $S_0 < 0$ to a final state of constant concentration where $S_\infty = 0$.

For the initial concentration

$$c(x, t = 0) = 1 + \cos 2\pi x \tag{3.38}$$

we find the solution

$$c(x, t) = 1 + e^{-Dt} \cos 2\pi x \tag{3.39}$$

For an initial concentration $c(x, t = 0) = \delta(x)$ we find for an unbounded domain

$$c(x, t) = \frac{1}{(4\pi Dt)^{1/2}} \exp\left\{-\frac{x^2}{4Dt}\right\} \tag{3.40}$$

The diffusion process is the prototype of an irreversible process.

3.5 Dynamical Systems

Environmental systems are very complex. They have an infinite number of degrees of freedom which interact in a highly nonlinear manner. They are open, i.e., in contacts with their surroundings. Even if we were able to formulate the exact equations of motion for such a system, it would be impossible to solve these equations. Approximations must be applied, often very drastic ones, in order to reduce the number of degrees of freedom to a manageable size. In other situations one formulates equations in an ad-hoc manner without detailed reference to some underlying physics. These procedures often yield a set of coupled ordinary differential equations

$$\dot{y}(t) = F(y(t)) \tag{3.41}$$

for a state vector $y(t) = (y_1(t), \ldots, y_N(t))$ of small dimension N. Systems described by (3.41) are called dynamical systems. The solution $y(t)$ describes a trajectory in the N-dimensional phase space spanned by y_1 to y_N. If the solutions for different initial conditions $y(t=0) = y_0$ are considered, one again has a mapping or flow in phase space:

$$y = y(y_0, t) \tag{3.42}$$

A typical example are the Lorenz equations:

$$\dot{x} = -\sigma x + \sigma y \tag{3.43a}$$
$$\dot{y} = rx - y - xz \tag{3.43b}$$
$$\dot{z} = xy - bz \tag{3.43c}$$

They are supposed to describe the essential features of a convective system where x is the circulation velocity, y the temperature difference between up and downward moving fluid particles, and z the deviation of the vertical temperature gradient from its equilibrium value. The constants $\sigma > 0$ and $b > 0$ describe properties of the fluid, and $r > 0$ is a control parameter which characterises the applied temperature difference ΔT that drives the convection. The Lorenz system is a system with three degrees of freedom. Its phase space is three-dimensional. The system has quadratic nonlinearities.

The analysis of dynamical systems is generally based on the geometric and topological properties of the mapping (3.42). If this mapping conserves phase space volume, the system is called conservative and it is very similar to the Hamiltonian systems considered in Sect. 3.2.4. If the mapping does not conserve phase space volume, it is called dissipative. If all trajectories approach a point, the system is very similar to the irreversible thermodynamic systems considered in Sect. 3.4. Between these extremes, dynamical systems show a wide range of intermediate behaviour.

Dynamical systems can show very irregular behavior. This irregular behavior is caused internally by the nonlinearities. The dynamic system is said to be chaotic when points in phase that are initially close to each other diverge from each other at a sufficiently fast (= exponential) rate.

A subspace of phase space is called an attractor, if all trajectories converge onto this subspace. Attractors can have very complex structures. They can have any dimension between 0 and N, including fractal dimensions.

A point y_0 that satisfies

$$F(y_0) = 0 \tag{3.44}$$

is called a fix point. If the eigenvalues of the stability matrix $\partial F_i / \partial y_j$ $(y = y_0)$ are all negative, then the fix point is stable and acts as an attractor for its immediate neighbourhood. If one of the eigenvalues is positive, the fix point is unstable. Systems with unstable fix points can become chaotic.

The analysis of the Lorenz system (3.43) shows that its behaviour is governed by the value of the control parameter r. The system has three fix points

$$x_1 = 0, y_1 = 0, z_1 = 0 \tag{3.45a}$$

$$x_{2,3} = \pm\sqrt{b(r-1)}, \ \ y_{2,3} = \pm\sqrt{b(r-1)}, \ \ z_{2,3} = r-1 \tag{3.45b}$$

The first point corresponds to a state of pure molecular heat conduction without any motion. The second fix point corresponds to a state of convective rolls. For $0 < r < 1$, the first point is stable. For $1 < r < r_c$ where

$$r_c = \sigma \frac{\sigma + b + 3}{\sigma - b - 1} \tag{3.46}$$

the second fix point becomes stable. For $r > r_c$ the system becomes chaotic with the famous "butterfly attractor", which has a dimension of 2.06.

The analysis of dynamical systems has shown that deterministic systems can generate irregular behaviour internally. It does not need to be imposed by irregular forcing or by randomness. This was a major insight.

A chaotic system is not predictable since we can never know its initial condition with ultimate precision. This raises the question of how to "verify" or "validate" its dynamics experimentally or observationally. The standard "verification" procedure which compares the theoretical prediction with the experimental or observed value fails.

Dynamical system equations are usually derived by ad-hoc procedures or by drastic approximations to basic physical equations. The Lorenz system is supposed to describe convective fluid flows either in the laboratory or in the atmosphere. These fluid flows have an infinite number of degrees of freedom. Because of these drastic reductions, it is not always clear to what extent the original system shares the properties of the reduced dynamical system.

3.6 Statistical Mechanics

Statistical mechanics attempts to derive the macroscopic phenomena and laws from the microscopic properties of the system. The prototype problem is the derivation of the thermodynamic properties of a substance from the properties of its constituent molecules. The derivations involve combinatoric and probabilistic arguments.

3.6.1 Combinatorics

Consider N different molecules $m1, ..., mN$. In how many ways can you distribute them into two boxes such that N_1 molecules are in the first box and $N_2 = N - N_1$ molecules are in the second box? This is a problem of combinatorics. Unordered selections of N_1 objects from N objects are called N_1-combinations and their number is given by

$$C(N_1) = \binom{N}{N_1} = \frac{N!}{N_1!(N - N_1)!} = \frac{N!}{N_1! N_2!} \tag{3.47}$$

If we define the microstate as the specification of which of the molecules $m1, ..., mN$ is in which box and the macrostate as the number of molecules in a box, then $C(N_1)$ is the number of microstates that lead to the same macrostate. If we assume that all microstates have the same probability, then the probability of the macrostate N_1 is given by

$$p(N_1) = \binom{N}{N_1}\left(\frac{1}{2}\right)^N \tag{3.48}$$

since

$$\sum_{N_1=0}^{N} \binom{N}{N_1} = 2^N .$$

This probability distribution is the binomial distribution where the probability of both events is 1/2. There is thus an alternative interpretation of the result (3.48), namely that (3.48) describes the probability of finding N_1 molecules in box 1 if one puts each of the N molecules with probability 1/2 in either box 1 or box 2. For large N, the binomial distribution can be approximated by the Gaussian distribution

$$p(N_1) \approx \frac{2}{\sqrt{2\pi N}} exp\left\{ -\frac{1}{2} \frac{(N_1 - N/2)^2}{N/4} \right\} \tag{3.49}$$

with mean $\mu = N/2$ and variance $\sigma^2 = N/4$. For large N we thus find that the distribution is sharply peaked at $N_1 = N/2$ with $\sigma/\mu = 1/(N)^{1/2}$.

Thermodynamics, i.e., diffusion theory, holds that the state with an equal number of molecules in both boxes is the thermodynamic equilibrium state, the state of maximum entropy. The above combinatoric/probabilistic arguments identify this thermodynamic equilibrium state as the state with the highest probability. The approach toward thermodynamic equilibrium is the approach from a less probable to a more probable state. Indeed, Boltzmann put this connection on a quantitative basis by postulating the following relation between the macroscopic entropy S and the microscopic probability p

$$S = k \ln p \tag{3.50}$$

where k is now called the Boltzmann constant. The higher the probability of a state, the higher is its entropy.

These connections between macroscopic laws and microscopic physics are, however, based on various assumptions. Foremost is the question why do all microstates have the same probability? Here one usually invokes the ergodic hypothesis. Firstly, it states that the microscopic system is ergodic, i.e., it "fills out" the phase space. Secondly, it states that the phase space is filled out uniformly. Time averages can be replaced by phase space averages where every microstate has the same probability. In most cases, this ergodic hypothesis cannot be proven explicitly or numerically but must be assumed. There is the additional problem of whether or not the ergodic hypothesis is independent of representation. If one chooses to describe the system by a different set of variables, will one get the same result? Equal probabilities in Cartesian coordinates differ from equal probabilities in spherical coordinates. Despite these fundamental problems, statistical mechanics has successfully been applied to a wide variety of problems.

3.6.2 H-theorem

Statistical mechanics also provides an explanation for the approach toward thermodynamic equilibrium. It is simply the approach to a more probable state. In a statistical mechanical context, the entropy S as given by (3.50) is usually denoted by H[1] and the statement that the entropy increases is called the H-theorem. Any proof of the H-theorem runs into a very basic problem. The laws that govern the microscopic system are reversible. How can one derive an irreversible macroscopic law from reversible microscopic dynamics? The probabilistic interpretation *seems* to be the way out; however, not quite as the following example will show.

Consider two bowls and N white marbles $w1,...,wN$ and N black marbles $b1,...,bN$. There are N marbles in each bowl. The dynamics is to take randomly one marble from each box and exchange them. The microstate is determined by which marble is in which bowl. The macrostate is that there are n_k white marbles in bowl 1 after k exchanges. The number n_k completely specifies the macrostate, because if there are n_k white marbles in bowl 1, then we know that there are $N - n_k$ black marbles in bowl 1 and $N - n_k$ white and n_k black marbles in bowl 2. The microscopic dynamics is stochastic, as opposed to deterministic, but it is reversible. The probability of going from one microstate A to another microstate B where two specific marbles are exchanged is given by

$$p^{AB} = \frac{1}{N^2} \tag{3.51}$$

and is the same as going from B to A

$$p^{BA} = \frac{1}{N^2} \tag{3.52}$$

Now consider the macrostate that there are n_k white marbles in bowl 1 after k exchanges. The number of possible microstates for this macrostate is

$$C(n_k) = \left(\frac{N!}{n_k!(N - n_k)!} \right)^2 \tag{3.53}$$

If all the microstates are assumed to have the same probability then the probability of the macrostate is given by

$$p(n_k) = \left(\frac{N!}{n_k!(N - n_k)!} \right)^2 \frac{(N!)^2}{(2N)!} \tag{3.54}$$

The entropy of this macrostate is according to Boltzmann's formula (3.50) with $k = 1$

---

[1] Note that in the original literature, H denotes minus the entropy.

$$H = ln\, p(n_k)$$

$$= ln\frac{(N!)^4}{(2N)!} - 2ln(N - n_k)! - 2ln\, n_k!$$

$$\approx ln\frac{(N!)^4}{(2N)!} - 2(N - n_k)ln(N - n_k) - 2n_k\, ln\, n_k \qquad (3.55)$$

where we have used Stirling's formula $ln\, N! \approx N \ln N$ which is a good approximation for large N. The maximum of H is obtained from the condition $\partial H/\partial n_k = 0$ and is reached for the macrostate

$$n_k = \frac{N}{2} \qquad (3.56)$$

This is the expected result.

Next, consider the probabilities to go from the macrostate n_k to the macrostate n_{k+1}, which are given by

$$p_+(n_k \rightarrow n_{k+1} = n_k + 1) = \frac{(N - n_k)^2}{N^2} \qquad (3.57a)$$

$$p_0(n_k \rightarrow n_{k+1} = n_k) = 2\frac{(N - n_k)n_k}{N^2} \qquad (3.57b)$$

$$p_-(n_k \rightarrow n_{k+1} = n_k - 1) = \frac{n_k^2}{N^2} \qquad (3.57c)$$

These probabilities can be rationalised as follows: In order to go from n_k to $n_{k+1} = n_k+1$ one has to take one black marble out of bowl 1. The probability for this is $(N - n_k)/N$, since there are $(N - n_k)$ black marbles in bowl 1. Simultaneously, one has to take one white marble out of bowl 2. The probability for this is $(N - n_k)/N$, since there are $(N - n_k)$ white marbles in bowl 2. The other probabilities follow by the same reasoning. Thus,

$$\frac{p_+}{p_-} = \frac{(N - n_k)^2}{n_k^2} \qquad (3.58)$$

If $n_k < N/2$, then $p_+/p_- > 1$. It is thus more probable that the system will approach the equilibrium value $n_k = N/2$ than to move away from it. For the entropy we find

$$H_{k+1} > H_k \qquad (3.59)$$

However, consider next where the system was most likely coming from. The probability that it came from $n_{k-1} = n_k +1$ is

$$p_{+0+} = p(n_{k-1} = n_k + 1)p_-(n_k + 1 \rightarrow n_k)p_+(n_k \rightarrow n_k + 1)$$

$$= \frac{(N!)^2}{(2N)!} \frac{(N!)^2}{\left[(N-n_k-1)!(n_k+1)!\right]^2} \frac{(n_k+1)^2}{N^2} \frac{(N-n_k)^2}{N^2} \tag{3.60}$$

Similarly, the probability that it came from $n_{k-1} = n_k - 1$ is

$$p_{-0+} = p(n_{k-1} = n_k - 1)p_+(n_k - 1 \rightarrow n_k)p_+(n_k \rightarrow n_k + 1)$$

$$= \frac{(N!)^2}{(2N)!} \frac{(N!)^2}{\left[(N-n_k+1)!(n_k-1)!\right]^2} \frac{(N-n_k+1)^2}{N^2} \frac{(N-n_k)^2}{N^2} \tag{3.61}$$

The ratio is

$$\frac{p_{+0+}}{p_{-0+}} = \frac{(N-n_k)^4}{n_k^2(N-n_k)^2} = \frac{(N-n_k)^2}{n_k^2} > 1 \tag{3.62}$$

if $n_k < N/2$. It is thus more probable that the system came from $n_{k-1} = n_k + 1$ than from $n_{k-1} = n_k - 1$. One can also prove that p_{+0+} is larger than any other path that goes through n_k. The state n_k is thus most likely an extremum. It came most likely from a state closer to equilibrium. It will move most likely to a state closer to equilibrium. For the entropy we find $H_{k-1} > H_k$ and $H_{k+1} > H_k$ or in differential form

$$\frac{dH}{dt} > 0 \quad \text{for } t > 0$$

$$\frac{dH}{dt} < 0 \quad \text{for } t < 0$$

The entropy does not increase monotonically, as in thermodynamics. The statistical mechanics approach implies that the system is fluctuating about its equilibrium state. The time symmetry is not broken by applying probabilistic arguments. One cannot derive irreversible macroscopic laws from reversible microscopic dynamics.

On the other hand, on philosophical grounds one might argue that one should apply probabilistic arguments only forward in time; to the future and not to the past. Only the future is open. The past is factual and is, in principle, known. When applying probabilistic arguments only forward in time, no inconsistencies arise. The approach toward thermodynamic equilibrium is an approach toward the most probable state.

3.7 Stochastic Processes

Probabilistic concepts can also be inserted directly into dynamical equations, transforming deterministic differential equations into stochastic differential equations. The random walk often serves as a starting point for this procedure.

3.7.1   Random Walk

Consider a particle moving on a discrete lattice ...,–2, –1, 0, 1, 2, ... It is released at time $t = 0$ at the position $m = 0$. At each time step it moves to the left or to the right with equal probability 1/2. What is the probability that it will be at position m after N time steps? To end up at m, the particle must make s move to the left and N-s move to the right where $s = (N-m)/2$. The probability for this to occur is again the binomial probability distribution

$$p(s) = \binom{N}{s} \left(\frac{1}{2}\right)^N \tag{3.63a}$$

or

$$p(s) \approx \frac{2}{\sqrt{2\pi N}} \exp\left\{-\frac{1}{2} \frac{(s-N/2)^2}{N/4}\right\} \tag{3.63b}$$

for large N as before. The probability for being at m is then given by

$$p(m) = p(s)\frac{ds}{dm} = \frac{1}{\sqrt{2\pi N}} \exp\left\{-\frac{1}{2}\frac{m^2}{N}\right\} \tag{3.64}$$

If we substitute $x = am$, where a is the distance between two adjacent lattice points and $t = N\tau$, where τ is the time interval between successive steps then we obtain

$$p(x,t) = p(m,N)\frac{dm}{dx} = \frac{1}{(4\pi Dt)^{1/2}} \exp\left\{-\frac{x^2}{4Dt}\right\} \tag{3.65}$$

where

$$D = \frac{a^2}{2\tau} \tag{3.66}$$

This solution for the probability density function is the same as the solution (3.40) for the concentration of the diffusion equation. We thus have discovered an underlying stochastic process for the macroscopic diffusion process. If individual particles perform a random walk, then their concentration is governed by a diffusion process with diffusion coefficient (3.66).

The transition from a stochastic process that is discrete in space and time to a process that is continuous can of course directly be made at the level of the governing equation, not only in the solution.

3.7.2 Autoregressive Process

Let us first consider space to be continuous. This leads to the autoregressive process

$$x_n = \alpha x_{n-1} + \eta_{n-1} \tag{3.67}$$

which describes how the "position" of a particle changes from time step $n - 1$ to time step n by a continuous random increment η_{n-1}. The coefficient α is a free parameter. The increments are generally chosen to be uncorrelated random variables each with a Gaussian distribution $N(0, \sigma^2)$ of zero mean and variance σ^2. Thus

$$< \eta_n >= 0 \tag{3.68a}$$

$$< \eta_n \eta_{n'} >= \begin{cases} \sigma^2 \text{ for } n = n' \\ 0 \text{ for } n \neq n' \end{cases} \tag{3.68b}$$

where angle brackets denote the expectation value.

For $\alpha = 1$, Eq. 3.67 describes a random walk. The variance of the random walk satisfies the recursion formula

$$< x_n x_n >=< x_{n-1} x_{n-1} > +\sigma^2 + 2 < x_{n-1} \eta_{n-1} >$$
$$=< x_{n-1} x_{n-1} > +\sigma^2 \tag{3.69}$$

since x_{n-1} only depends on η_{n-k} with $k > 1$. At each time step the variance thus increases by σ^2. The random walk is nonstationary with ever increasing variance

$$\lim_{n \to \infty} < x_n x_n >= n\sigma^2 \tag{3.70}$$

For $|\alpha| < 1$, the autoregressive process becomes stationary. Without any random forcing η, the system will approach the limit $x = 0$. The random forcing keeps the system "alive." However, being alive only means fluctuations about the mean state $<x> = 0$.

Generalisations of the autoregressive process (3.67) are the ARMA (p,q) processes

$$x_n = \sum_{k=1}^{p} \alpha_k x_{n-k} + \sum_{k=1}^{q} \beta_k \eta_{n-k} \tag{3.71}$$

where α_k and β_k are constants. ARMA stands for autoregressive moving average. The process (3.67) is an ARMA(1,0)=AR(1) process.

3.7.3 Langevin Equation

Time can also be made continuous. One then arrives at the Langevin equation

$$\frac{dx(t)}{dt} = -\gamma x(t) + \zeta(t) \tag{3.72}$$

where γ is a damping constant and where $\zeta(t)$ is now a random Gaussian process with the properties

$$\langle \zeta(t) \rangle = 0 \tag{3.73a}$$
$$\langle \zeta(t)\zeta(t') \rangle = \sigma^2 \delta(t-t') \tag{3.73b}$$

This process is called white noise and denoted by $WN(0, \sigma^2)$. The white noise process has infinite variance. This causes certain complications. One therefore introduces the Wiener process

$$W(t) = \int^t dt' \zeta(t') \tag{3.74}$$

which is the random walk generated by white noise. It is a nonstationary process whose variance grows linearly in time

$$\langle W(t)W(t) \rangle = \sigma^2 t \tag{3.75}$$

The increment of the Wiener process

$$dW(t) = \int_t^{t+dt} dt' \zeta(t') \tag{3.76}$$

has the properties

$$\langle dW(t) \rangle = 0 \tag{3.77a}$$
$$\langle dW(t)dW(t') \rangle = 0 \text{ for } t \neq t' \tag{3.77b}$$
$$\langle dW(t)dW(t) \rangle = \sigma^2 dt \tag{3.77c}$$

and can be used to rewrite the Langevin equation as

$$dx(t) = -yx(t) + dW(t) \tag{3.78}$$

This form most clearly shows the difference between stochastic and deterministic forcing. If instead of the white noise forcing term $\zeta(t)$ we had a deterministic forcing term $f(t)$ in the Langevin equation (3.72), then the increment in (3.78) would be $f(t)dt$, whereas it is $dW(t) \sim \sqrt{dt}$ for white noise forcing. The deterministic increment is proportional to dt. The stochastic increment is proportional to the square root of dt. This difference reflects the fact that the white noise process is so irregular, that standard estimates fail for the integral (3.76). This irregularity also causes the variance of the random walk to grow more slowly, proportional to t (or n) as we found in (3.75) (or (3.70)) and not proportional to t^2 as it would for deterministic forcing. The irregular structure of white noise forcing also requires a new calculus, as we see next.

3.7.4 Stochastic Differential Equations

A straightforward generalisation of the Langevin equation is

$$dx(t) = A(x(t),t)dt + B(x(t),t)dW(t) \tag{3.79}$$

where A und B are arbitrary functions. If the function B depends on $x(t)$, then the noise becomes multiplicative, as opposed to additive. Equation (3.79) is the prototype of an ordinary stochastic differential equation. Its interpretation is, however, not unique. This can be seen as follows. Integrate (3.79) from t to $t+dt$. One obtains

$$dx(t) = \int_t^{t+dt} A(x(t'))dt' + \int_t^{t+dt} B(x(t'))dW(t') \tag{3.80}$$

where we have neglected the irrelevant explicit time dependence of the functions A and B. If we substitute the initial value $x(t)$ for $x(t')$ in the integrands, then we recover (3.79). However, if we substitute a weighted average between the initial value and the end value

$$\hat{x}(t) = (1-\alpha)x(t) + \alpha x(t+dt)$$
$$= x(t) + \alpha dx(t) \tag{3.81}$$

then we obtain

$$dx(t) = A(x(t) + \alpha dx(t))dt + B(x(t) + \alpha dx(t))dW(t), \tag{3.82}$$

which is an implicit equation for $dx(t)$. The lowest order contributions to $dx(t)$ are proportional to $\sqrt{dt}$. Taylor expansion of A gives a first correction term proportional to $\alpha \, dx(t) \, dt \sim \alpha \, (dt)3/2$, which is of higher order than dt. The value of α hence does not matter in the first term. However, the value of α does matter in the second term, since the first correction term is proportional to $\alpha \, dx \, (t)dW(t) \sim \alpha \, dt$. Different choices of α obviously lead to different results. It should also be obvious that the rules of calculus need to be modified because of the stochastic forcing term $dW(t) \sim \sqrt{dt}$. The Ito calculus assumes $\alpha = 0$. The Stratonovich calculus assumes $\alpha = 1/2$.

Differential equations can be driven by other than white noise processes, such as birth and death and branching processes. Stochastic terms can also be added to partial differential equations. Solutions of stochastic differential equations show behavior that is often impossible (or at least difficult) to achieve with deterministic equations. Patchiness, i.e., the clustering into disconnected communities, can for example be obtained relatively easily by combining a random walk with a random birth and death process. Overall, stochastic differential equations and the resulting stochastic processes offer a tool to describe and understand complex environmental systems. They ascribe irregular behavior to stochastic elements in the system in contrast to dynamical systems where irregular behavior is caused by nonlinearities in the system.

3.8 Discussion

Environmental modelers must formulate dynamical equations that represent the "physics" of the system. This is not as straightforward a task as it may seem. There are a variety of physical theories that are based on different idealised circumstances. The physical system can consist of discrete particles or of continuous fields. The dynamical evolution can be deterministic or stochastic, reversible or dissipative, and predictable or chaotic. Which of these concepts is most adequate to the environmental system at hand must be decided by the modeler. Some concepts are mutually exclusive. We specifically tried to contrast reversible versus dissipative systems and deterministic versus stochastic systems.

For a reversible Hamiltonian system, the flow in phase space is non-divergent, whereas for a dissipative system the flow contracts onto an attractor of lower dimension. Both systems can show highly irregular behavior. One can have mixing flows for Hamiltonian systems and attractors with fractal dimensions for dissipative systems. Even reversible Hamiltonian systems exhibit a certain kind of irreversibility. Coarse information about the system degrades in time in mixing flows. This does not contradict the reversibility of the underlying dynamics. What is degrading here is the information. If one has complete information at the initial time, then one has complete information at all times. However, if one has incomplete information about the system initially, then one has even more incomplete information about it later in time. This property led to the idea that probability concepts can bridge the gap from reversible microscopic dynamics to irreversible macroscopic laws. Indeed, one can show that a reversible microscopic system evolves from an unlikely state to a more likely state. This can be interpreted as the irreversible evolution towards thermodynamic equilibrium. However, the same reasoning will also show that the reversible microscopic system came from a more likely state. Thus the introduction of probability concepts does not break the time symmetry of the microscopic dynamics. One cannot obtain true irreversible behavior. However, one might argue on philosophical grounds that one should apply probabilistic concepts only forward in time and not to the past, since the past is factual and in principle known. Then the probabilistic interpretation gives an explanation for the approach toward thermodynamic equilibrium.

A second major issue is the distinction between deterministic and stochastic systems. This is quite a contentious issue based on philosophical convictions that the world is or is not random. Although deterministic and stochastic systems are quite distinct in principle, the distinction becomes blurred in practice. It is quite difficult to distinguish irregular deterministic behaviour from irregular stochastic behaviour. This is implicitly acknowledged by everyone who applies statistical analysis techniques to the output of numerical models that are deterministic. Though the distinction between deterministic and stochastic systems might be a grand philosophical issue it is perhaps much less of an issue in practical applications. Stochastic processes should be introduced when we are uncertain about the exact nature of the dynamics. Randomness is then a statement about our knowledge and not about the true nature of the world. This is also the appropriate attitude towards dynamical modeling. The dynamical equations represent our subjective information about the system and not the true physics of the system.

Acknowledgments

The author would like to thank the organisers of the Fall School for the opportunity to participate in a most stimulating and enjoyable learning experience and Diane Henderson and Sharon Sakamoto for their help in the preparation of the manuscript.

Chapter 4
Process-oriented Models in Physical Oceanography

by Aike Beckmann

Abstract

Process-oriented models are widely used in oceanography to understand physical mechanisms and their parameter dependence, to dynamically interpret observations and to determine the role of individual processes relative to each other. This paper gives an overview of concepts, strategies and areas of application of such models in the field of physical oceanography and related interdisciplinary marine research. Both strengths and limitations of the various approaches are described. Selected examples will be presented for the whole suite of available models, covering the range from linear wave models to fully three-dimensional non-hydrostatic deep convection models. The additional complexity introduced by the use of numerical approximations is emphasised.

4.1 Introduction

The basis for physical oceanography are the so-called non-hydrostatic primitive equations (NHPE) which consist of six prognostic conservation equations and one diagnostic relation and are an extension of the Navier-Stokes equations for a stratified, rotating fluid (see, e.g., Gill, 1982):

momentum: $\quad \dfrac{d\vec{v}\rho}{dt} + 2\vec{\Omega} \times \rho\vec{v} \;=\; -\nabla\rho + \rho\nabla\Phi_G + \rho\nabla\Phi_T + F^{\vec{v}} + D^{\vec{v}}$ $\quad$ (4.1)

mass: $\quad \dfrac{d\rho}{dt} \;=\; 0$ $\hspace{6cm}$ (4.2)

internal energy: $\quad \dfrac{dT}{dt} \;=\; F^T + D^T$ $\hspace{3.5cm}$ (4.3)

partial mass: $\quad \dfrac{dS}{dt} \;=\; F^S + D^S$ $\hspace{3.5cm}$ (4.4)

equation of state: $\;\rho \;=\; \rho(S,T,p)$ $\hspace{3.5cm}$ (4.5)

where the total derivative d/dt includes both individual and advective rates of change. Here, u, v, w denote the three components of the velocity vector $\vec{v}$, T is the potential temperature, S the salinity and ρ the density of the fluid. The gravitational and tidal potentials are Φ_G and Φ_T, respectively; Ω is the angular velocity of the Earth's rotation. The *F* and *D* terms denote forcing and dissipation. It should be noted that the dynamical core of this system is very similar to the equations used in dynamic meteorology (where salinity is replaced by relative humidity).

In combination with a set of suitable boundary conditions (specified fluxes across the boundaries of the ocean), the above system describes the large variety of oceanic processes and phenomena, spanning the range from the largest basin wide gyres to molecular processes like double diffusion. Listed roughly according to spatial scale, we have

- basin-wide gyres, meridional overturning motion;
- tides;
- planetary waves, topographic waves;
- western boundary currents;
- mixed layer dynamics, ventilation, and convection;
- current instabilities, eddies, fronts;
- mesoscale topographic effects: throughflows, overflows;
- surface waves, internal waves;
- double diffusion.

Complications to solve these equations arise from three different conditions: the nonlinearities in the above equations, the irregular geometry of the ocean (with fractal coastlines, rough and steep topography) and a forcing, which is highly variable in both space and time (and generally comprises momentum, heat and fresh water fluxes as well as tides).

4.1.1   Philosophy of Process Models

Based on the *a priori* assumption that individual processes described by the NHPE system can be studied successfully in isolation, a large number of process-oriented models has been developed. The goal of such a process model is to find a solution of a simplified or approximated system, addressing one (or few) relevant aspects. This has been done with analytical means, within laboratory experiments or by numerical integration. The latter two are especially valuable to extend analytical process models into the nonlinear, complex geometry regime.

To further define the philosophy in process-oriented modeling, we note:

- a *process model* investigates the kinematics and/or dynamics of individual physical processes in isolation or combination (deliberately omitting other, possibly important processes) using an idealised configuration to determine the mechanisms (and role) of these processes, and their parameter dependence. This includes both conceptual, analytical and numerical models.
  The alternative approach can be defined as
- a *simulation model* aims to forecast/hindcast the state (and evolution) of the ocean (circulation and water mass distribution) as closely as possible. These simulations are usually called General Circulation Models (GCMs), and have to be numerical. GCMs need not be global; even a regional study may consider

the full system of equations and a quasi-realistic approximation to the geometry and forcing.

Somewhat in between, there is another frequently used approach:

- a *parameter study* is a systematic variation of physical and/or numerical parameters of a given problem to determine fundamental dependencies and to classify the phenomenology. The definition of dynamical regimes is the result of such a study. A parameter study is an integral part of any process study, and is often applied to simulations as well, especially in connection with uncertainties in forcing and numerical algorithms[1].

4.1.2 Process Model Strategies

The methods to develop a process model involve *simplification*, *idealisation* and *abstraction* (i.e., finding a suitable simplified model configuration and identifying the minimum number of ingredients for an observed phenomenon or hypothesised mechanism), and can be categorised as follows:

- reduction of dimension or degrees of freedom (the limitation to three, two or one dimensions (e.g., steady state solutions, vertically integrated equations), a rigorous scale selection or separation, "box"- or low order models);
- simplification of the dynamics, usually by elimination of certain processes through systematic higher order expansion or explicit filtering (widely used examples include linearisation (e.g., the restriction to infinitesimal amplitudes or the definition of a suitable reference state), the use of a single state variable (SSV), which can be interpreted as a linear approximation to the equation of state, other formal scaling procedures (applying the Boussinesq, hydrostatic, traditional, thin shell and geostrophic approximations), which leads, e.g., to the hydrostatic primitive equations (HPE), and quasigeostrophy (QG), linear balance (LBE) or reduced gravity (RG) models. See Appendix.);
- simplification of ambient fields, geometry and/or forcing (flat bottom, approximation of the spherical Earth by a plane with constant (fo) or linearly varying (β) Coriolis parameter ($f = 2 \, \Omega \sin\phi$), straight coastlines, periodic channel configurations; simple parameterisations of the net effects of internal and external mechanisms (e.g., wind or tidal forcing)).

To make a critical evaluation of the range of validity possible, each approximation and simplification used to derive a process model needs to be explicitly listed.

---

[1] When considering a *numerical* process model, we implicitly assume that the process under consideration is resolved by the model grid. If this requirement is not met, the study does not directly address the physics of the process, but rather its representation in a discretised system. This can be of interest as well (e.g., coarse resolution climate models need to address the issue of unresolved physics very carefully), but has an entirely different character. See also Sect. 6.3.2.

In particular, one should be aware that some of the methods eliminate certain processes altogether, while others distort them[2].

4.1.3 Parameter Space

Often, a small number of non-dimensional parameters can be specified to describe the problem and to explore general parameter dependencies. Some generalisation of the results can be obtained.

The scales and parameters of the physical problem are related to the geometry (vertical scale h, horizontal scale L), its environment (water depth H, Coriolis parameter f, stability frequency N), and the velocity amplitude U. From these, a number of non-dimensional parameters can be obtained, which span the parameter space for the process model. Even in a very idealised setting, there are usually several independently specifiable quantities. The most important for a rotating stratified fluid are the *Rossby number*

$$R_0 = \frac{U}{fL}$$

which characterises the relative versus planetary vorticity; the *Burger number*

$$S = \frac{NH}{fL}$$

which relates the strengths of stratification and rotation, and the *Froude number*

$$F = \frac{U}{NH},$$

which compares flow and gravity wave speeds. In addition, the *Ekman and Reynolds numbers*

$$E = \frac{A_v}{fH^2}$$

and

$$Re = \frac{uL}{A_h}$$

---

[2] For example, an f-plane approximation eliminates Rossby waves, a flat bottom eliminates all topographic waves. On the other hand, the QG system causes an unrealistic symmetry between cyclones and anticyclones. See also Sect. 6.2.1.1, 6.2.2.3 and 6.2.2.7.

give estimates of the influence of friction. Additional parameters may appear for some processes; for example, the frequency ratio

$$v = \frac{\omega}{f}$$

for periodic processes, the vertical aspect ratio

$$\delta = \frac{h}{H}$$

for baroclinic processes, or the topographic steepness

$$\alpha = \frac{h}{L}$$

for topographically influenced processes.

4.2 Examples

This section is intended to present an overview of widely used concepts, all of which are in important areas of physical oceanography, and to show the evolution of methods and knowledge in the last twenty years. Successes will be highlighted; limitations and remaining open questions will be addressed.

4.2.1 Linear and Linearised Models

Standard manipulations of the NHPE system include the Boussinesq, hydrostatic, traditional, thin shell and incompressibility approximations. These are derived by scaling arguments and systematic filtering (see Appendix C). A further assumption, sufficiently small amplitudes of all prognostic variables, allows for the omission of all nonlinear terms.

4.2.1.1 Linear Waves

Waves constitute the ocean's response to perturbations; they are the mechanism of *gravitational and geostrophic adjustment*. A fundamental approach is therefore to look for *wavelike solutions* in a Cartesian framework. Such periodic analytical solutions of simplified equations of motion have been known even in the last century. They can be seen as the first "models" employed in oceanography.

The "model" in this case is the assumption of sinusoidal spatial and temporal dependence on the hydrodynamic variables in the form

$$
\begin{pmatrix} u \\ v \\ w \\ p \\ \rho \end{pmatrix} \sim \exp\left[i\left(kx + ly + mz - \omega t\right)\right], \tag{4.6}
$$

where k, l, m are wavenumbers in x, y, z direction and ω is the frequency. The most fundamental considerations in this approach focus on *free waves* and ignore generating mechanisms (e.g., resonance phenomena), frictional effects, as well as lateral boundaries. These limitations can be relaxed to include a linear damping, the effects of rotation and lateral boundaries (e.g., reflection), variable stratification and topography, and an approximate treatment of wave-wave interactions. A detailed overview can be found in LeBlond and Mysak (1987). The result of such a model is a dispersion relation (describing the relation between wave frequency and wavelength, depending on environmental parameters like rotation and strength of the stratification), as well as knowledge of phase speed and energy propagation.

Depending on the specific approximations, different types of waves can be identified:

Inertia-gravity waves. The classical wave theory begins with surface gravity waves, which are a special form of propagating boundary layer perturbation (Fig. 4.1a). In the presence of rotation and a lateral wall, these waves become trapped to the sidewall and are then called *Kelvin waves* (see Fig. 4.1b). These Kelvin waves can be used to construct a solution for abruptly varying topography (relative to the wavelength). In such a model, two waves in water of constant but different depth have to be matched at the topographic step. The resulting solution is called a *double Kelvin wave*.

For a stratified medium, internal waves are also possible: While surface gravity waves decay linearly or exponentially with depth, the vertical structure of internal

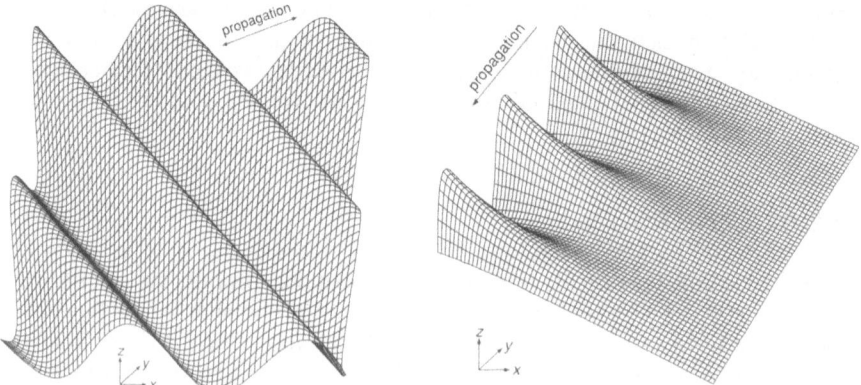

Fig. 4.1. Horizontal structure of (a) a plane wave, (b) a boundary wave (Kelvin wave). While plane waves propagation is isotropic, Kelvin waves are anisotropic with the lateral boundary to the right (on the northern hemisphere)

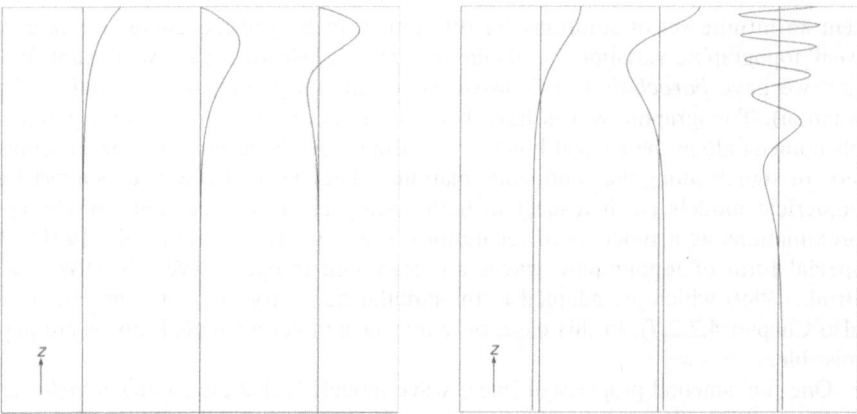

Fig. 4.2. Vertical modal structure of linear waves for a typical oceanic density stratification: (a) vertical velocity profiles for surface and internal waves, (b) horizontal velocity profiles for baroclinic planetary waves

waves has to be computed from an eigenvalue problem[3]. The resulting *modes* have internal maxima in the vertical velocity (Fig. 4.2a).

Vorticity waves. Waves of the second kind do not depend on gravity as restoring force but exist due to the variation of ambient potential vorticity (e.g., the planetary vorticity $f = 2\ \Omega\sin\phi$. Their analytical treatment is based on the *quasi-geostrophic potential vorticity equation* (see Appendix 4.5.2 for details of the derivation). If a linear approximation to the dependence of f on latitude is assumed ($f = f_o + \beta y$), these waves are called *Rossby waves.*

In the horizontal, a typical wavelike structure is assumed; in the vertical both barotropic (vertically unsheared) and baroclinic (sheared) waves can be found; the latter results again from a separation approach (see Fig. 4.2b). In contrast to inertia-gravity waves, the phase propagation of planetary waves is always westward.

Admittedly, these vertical modes are a highly idealised view of the ocean. Strictly speaking, variable bottom topography will prohibit modal solutions. In practice, however, this concept works well even for a smoothly varying bottom. Thus, the modal structure is one of the examples of how simplified concepts are successfully used for a much broader range of circumstances.

In analogy to planetary waves, vortex stretching can act as a restoring force in areas of large scale variations in topography. The analogy between topographic and planetary waves can be shown if the variable topography is assumed to be slowly varying (with respect to the wavelength) and of exponential shape $h = h_o$ $exp\ (-\alpha y)$. In this case, $f_o\ \alpha$ replaces β. Although the form of the topography is highly idealised, this model clearly demonstrates the correspondence between planetary and topographic waves.

As a result, there is an additional class of freely propagating subinertial-frequency waves in the ocean: so-called *coastal trapped waves.* They also repre-

---

[3] The separation of horizontal and vertical dependence is only possible under the assumption of a flat bottom.

sent an infinite set of solutions for different wavelength/frequencies. In case of weak topographic variation we obtain *barotropic shelf waves*; for weak stratification we have *baroclinic Kelvin waves* (i.e., internal gravity waves modified by rotation). Topographic waves have been used successfully to interpret periodic phenomena along the coastal boundaries of the world's ocean (e.g., the transmission of signals along the continental margins). They have also served as a test for numerical models (with respect to both coding errors and accuracy of the approximations as a function of resolution), see, e.g., Haidvogel et al. (1991). A special form of topographic waves are seamount trapped waves (SMTWs, see Brink, 1989) which are adapted to the circular geometry of submarine hills (see also Chapter 4.2.2.7). In this case, only integer number azimuthal modes are permissible.

One fundamental property of linear wave models is that the approach yields an infinite set of solutions, and any field can be decomposed into a set of waves. Whether appropriate or not, even a highly nonlinear eddy can be constructed by superposition of linear waves. But despite this apparent universality, the forecast skills are not always sufficient, as the ocean at the mesoscale is more nonlinear than linear.

4.2.1.2 Ekman Theory

Essential dynamics of the vertical boundary layers of the ocean have been derived by a consideration of Coriolis and vertical mixing terms only. This has led to Ekman theory (Ekman, 1905), which predicts the depth of the wind mixed boundary layer of the ocean, the vertical shear within this layer, and the net mass transport due to wind. The surface current is directed at 45° to the wind, and deeper flow rotates clockwise on the northern hemisphere (Ekman spiral; see Fig. 4.3). A simi-

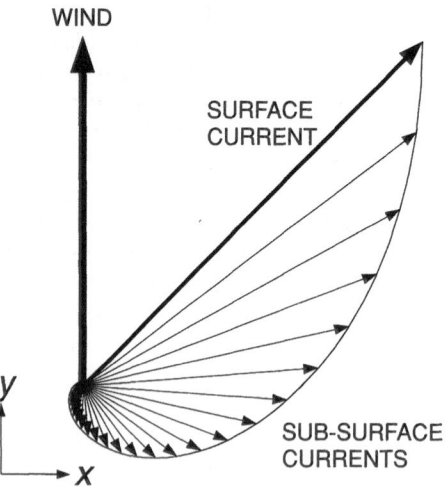

Fig. 4.3. Turning of the horizontal flow with depth (Ekman spiral)

lar, counterrotating spiral exists at the bottom of the ocean. The inherent assumptions are of this model: infinitely deep ocean, no lateral boundaries or gradients, constant vertical mixing, constant Coriolis parameter, a homogeneous ocean, and a time-independent wind[4].

Although these simplifications are hardly ever met in the real world, this model offers a unique way of explaining the large scale wind-driven circulation of the oceans. According to this extremely simple model, the vertically integrated transport within the surface (Ekman) layer is always to the right of the wind (on the northern hemisphere), and this will drive the large scale gyres in the interior of the ocean. Horizontal inhomogeneities in the Ekman currents induced by either a non-uniform wind field or coastal boundaries will lead to flow convergences or divergences; these cause vertical movements, the Ekman pumping/suction (see also 4.2.1.4). One of the strengths of this model lies in the insensitivity of this main result on the (unknown) value of the vertical mixing coefficient.

4.2.1.3 Westward Intensification

Based on a set of vertically integrated equations (see Appendix C), and ignoring thermohaline forcing, the linear steady state wind-driven circulation on a β-plane can be calculated analytically, if a purely zonal, sinusoidal wind field; a rectangular basin and a flat bottom are assumed.

The first of a long series of models of the large-scale circulation were developed on this basis by Stommel (1948), and Munk (1950). These models were able to explain the westward intensification (i.e., the occurrence of strong boundary currents along the western rim of the oceans) as the result of the β-effect.

Different frictional parameterisations yield slightly different forms of the resulting barotropic transport streamfunction. In particular, a higher order term can explain the recirculation close to the western boundary. Figure 4.4 shows such an example of the Munk solution (with biharmonic mixing) in a quadratic basin, thought to represent a subtropical gyre on the northern hemisphere.

4.2.1.4 Thermocline Ventilation

One of the fundamental goals of oceanography is to explain the observed density structure of the ocean. In the early 1980s, a linear analytical theory was put forward to explain the maintenance of the permanent thermocline (the sharp transition between the oceanic warm and cold water spheres) in the ocean. Using symp-

---

[4] Later, time-dependence of the wind had been included and showed that the transient response is governed by inertial waves.

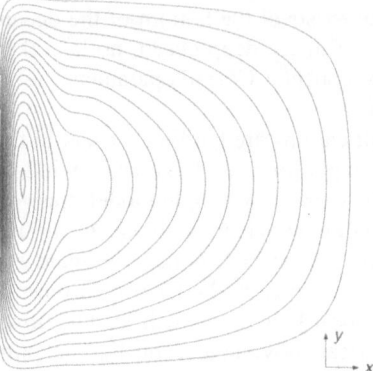

Fig. 4.4. Steady state solution to the Munk problem in a quadratic ocean with sinusoidal zonal wind stress. Shown are isolines of the mass transport streamfunction. The circulation is clockwise

lifying assumptions, analytical solutions could be obtained that describe some of the aspects of mid-latitude subsurface circulation and water mass distribution.

Based on potential vorticity conservation (see Appendix), models for the ventilation of the thermocline and the subduction process were developed (Luyten et al. 1983). The model assumes inviscid layers of homogeneous fluid (see Fig. 4.5); under these assumptions, it can be shown that the formation of water masses in subtropical gyres is confined to regions where the Ekman pumping velocity at the base of the mixed layer (see 4.2.1.2) is downward. Gradual downward and westward motion of the newly formed water masses leads to a slow ventilation of the deeper layers of the ocean. As a direct consequence, *shadow zones* develop in the

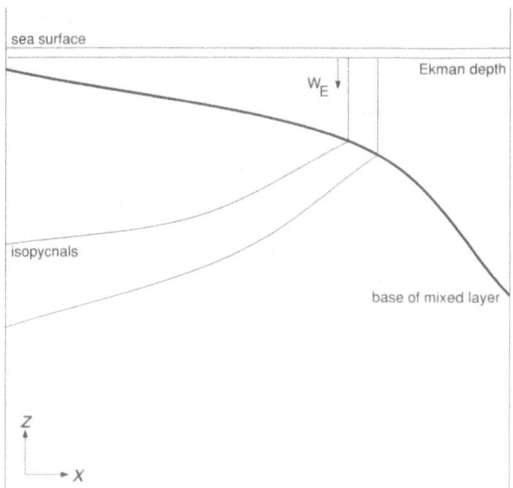

Fig. 4.5. Subduction of a layer of homogeneous density at the base of the mixed layer

eastern subtropical oceans that are not ventilated. Another result of these models was the existence of an area of homogenised potential vorticity within the subtropical gyre. These models were extended and refined in later years; Luyten and Stommel (1986) included buoyancy forcing and nonlinearity of the flow, which leads to a cyclonic circulation within the shadow zone. The validity of theses models has been shown in climatological data sets and numerical models of the basin-wide wind-driven and thermohaline circulation. For a more detailed overview on this subject, the reader is referred to Pedlosky (1996).

4.2.1.5 Linearised Barotropic/Baroclinic Instability

On a smaller scale, vorticity dynamics plays a fundamental role in generating wave-like disturbances in the ocean. In analogy to the atmosphere, current instability models were applied to oceanic flows. These flow instability models are a natural extension to the linear wave models: normal mode solutions are sought to a *linearised problem* that includes a (fixed in time and space) background flow (which itself needs to be a solution to the governing equations).

This approach forms the basis for models of barotropic/baroclinic instability, which have been used to obtain an understanding of the growth of disturbances and cyclogenesis in the atmosphere and the oceans. The quasigeostrophic potential vorticity equation (see Appendix B) is linearised around by setting state:

$$u = u_o - \frac{\partial \psi}{\partial y} , \quad v = \frac{\partial \psi}{\partial x}.$$

The resulting equation is then

$$\frac{\partial}{\partial t}\left(\nabla^2 - \lambda^{-2}\right)\psi + u_o \frac{\partial}{\partial x}\left(\nabla^2 - \lambda^{-2}\right)\psi + \beta\psi_z = 0 \tag{4.7}$$

Here, the background flow $u_o(y,z)$ is any function of the vertical and cross-stream coordinate. The system can be solved for the vertical and cross-stream dependence by assuming wavelike solutions in the along-stream direction.

In the framework of these models, it was possible to derive *necessary conditions for instability*, namely that the gradient of the ambient potential vorticity has to change sign. In addition, typical growth rates could be computed, which give a rough estimate of the time scales involved in these instability processes. The maximum unstable waves were found to be close to $k = R_D^{-1}$, a feature observed repeatedly in the ocean (Stammer and Böning 1992). Detailed investigations of the vertical structure of such unstable modes were performed by Beckmann (1988).

One of the inherent shortcomings of this class of models is that any feed-back between the background state and the wave perturbations is excluded. Therefore they cannot give a realistic prediction of the long-term evolution of unstable currents. Since the early 1990s, these models are replaced by fully nonlinear models that include the effects of the waves on the mean flow (see also Sect. 4.2.2.3).

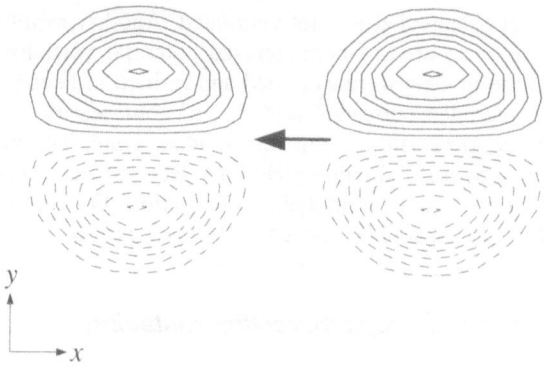

Fig. 4.6. Soliton solution to the nonlinear quasigeostrophic vorticity equation. The dipole structure (here shown in the streamfunction field) moves on a β-plane without change of shape

4.2.2 Nonlinear Models

Consideration of nonlinear effects usually requires laboratory or numerical models. Basis for these numerical realisations of ocean processes are two- or three-dimensional simplified versions of the NHPE system. The most widely used are (with increasing level of complexity): the shallow water equations (SWE), the quasigeostrophic equations (QG) and the hydrostatic primitive equations (HPE), (see the Appendix).

However, before we take a look at examples of these numerical process models, a special analytical solution to the fully nonlinear equations needs to be mentioned.

4.2.2.1 Nonlinear Analytical Solutions

The most prominent example of an analytical solution to the nonlinear barotropic (or quasigeostrophic) equation is a solitary wave, which is characterised by a complete balance between advective and dispersive terms. These structures propagate but do not change their shape with time (see Fig. 4.6). Asymptotic analytical solutions for such an equatorial Rossby soliton were given by Boyd (1985, 1989), using non-dimensionalised SWEs (see Appendix C).

These structures are relevant for physical oceanography, as they represent robust features (e.g., dipoles) that can survive for extended periods even in a turbulent eddy field. Their evolution and interactions with ambient fields have been studied in the context of geostrophic geophysical turbulence.

4.2.2.2 Gulf Stream Separation

The problem of *Gulf Stream separation* has been puzzling oceanographers for a long time. Linear barotropic theory (see 4.2.1.3) predicts that a western boundary current should stay at the boundary. The Gulf Stream, however, leaves the coast at Cape Hatteras, where it becomes a free meandering jet. The specific question is: Why does the Gulf Stream leave the coast?

Several hypotheses for Gulf Stream separation (Dengg et al. 1996) have been put forward:

- direct wind forcing separation;
- separation by detachment (assuming a dynamically forced surfacing of isopycnals);
- "vorticity crisis" (the model considers dynamical constraints on advection);
- inertial overshooting (in addition to advection, the shape of coastline is assumed to play a major role);
- baroclinic effects (the deep western boundary current is crossing the Gulf Stream in this area and might contribute to the separation of the Gulf Stream from the coast); and
- topographic effects (various, topography related mechanisms have been considered, like the depth change at the continental shelf break and the joint effect of baroclinicity and relief (JEBAR)).

For each of these hypotheses, a separate process model can be built, and over the years, this aspect was investigated with a large number of different models. Each of these has focused on a particular aspect of the dynamics and tried to explain the observed path of the Gulf Stream west of Cape Hatteras within the model framework. For example, diagnostic models have been employed to evaluate the effects of JEBAR (Mellor et al. 1982). Other models ignore baroclinic contributions and focus on the wind-driven circulation. For example, Fig. 4.7. shows a model that is used to investigate the effects of curved coastline, boundary conditions and nonlinearity on the path of the Gulf Stream in the northwest Atlantic Ocean (see also Dengg 1993). This model assumes that oceanic flows near the surface are mostly two-dimensional, and that density and topography effects are small, or cancel each other to a certain degree[5]. Reduced gravity (see Appendix D) and isopycnal models have been used to explore the effects of surfacing ("outcropping") of isopycnals (Parsons 1969, Chassignet and Bleck 1993; Chassignet et al. 1995).

Of course, most of these models can be "tuned" to the point where a realistically looking Gulf Stream path is obtained. It is therefore difficult to attribute the path of the western boundary current in the North Atlantic unequivocally to just one process. It may be fair to conclude that several of these mechanisms constructively act together[6].

---

[5] In fact, it can be argued that stratification will shield the near-surface flow from most of the effects of topography.

[6] The problem of Gulf Stream separation is still unresolved. From the point of numerical modeling, sufficient resolution (about 10 km horizontally and 10 m vertically near the surface) pro-

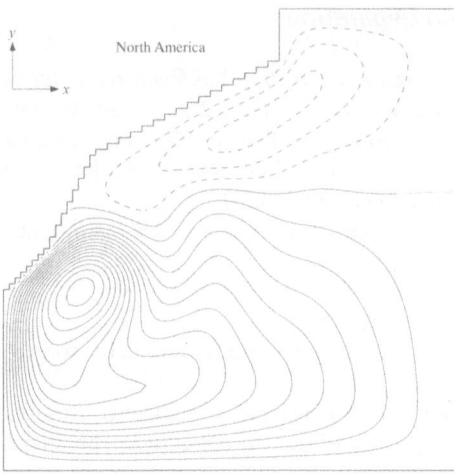

Fig. 4.7. A simplified configuration (zonal wind, idealised coastline) for the study of the Gulf Stream separation phenomenon. Isolines show the steady state mass transport streamfunction (solid lines: clockwise; dashed lines: anti-clockwise circulation. Strong nonlinearity in combination with a counter-rotating subpolar gyre can lead to separation

4.2.2.3 Frontal Instability, Eddy Generation

Satellite pictures of the ocean reveal the highly variable nature of the near-surface temperature fields. There is an abundance of eddies with a continuous spectrum of scales that fill the ocean. The generation mechanisms of these eddies is one of the important questions of large-scale oceanography. High frequency variability of the wind field can be one source of oceanic variability. But as enhanced levels of eddy kinetic energy are observed along strong current bands, it is assumed that these are one of the main sources.

Models to address this question are a direct extension of the linearised wave instability models presented in Sect. 4.2.1.5. They use the fully nonlinear QG equation (see Appendix B), or, more recently the HPE system and are solved numerically.

A typical configuration for the study of eddy generation mechanisms is a periodic channel with a zonal flow band (designed after the North Atlantic Current, the Azores Current or the Antarctic Circumpolar Current) that is perturbed by either random fluctuations or a localised disturbance. The goals are to reproduce the observed phenomenology (meandering jets, eddy detachment) and scales (temporal and spatial), and to quantify cross-frontal transports of heat and other tracers.

Such a mid-latitude mesoscale instability model was developed within the QG framework and applied to the Azores Current (Beckmann 1988). An example for a

---

duces a quite realistic Gulf Stream path. However, it remains unclear which effect or which combination of effects is responsible.

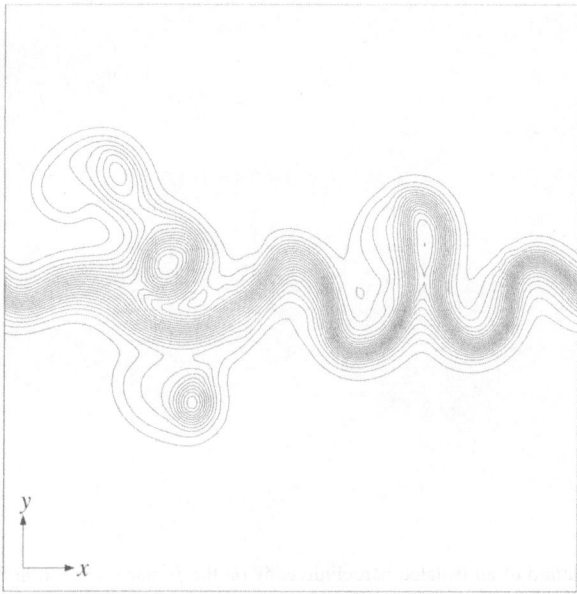

Fig. 4.8. Instability of a density front with wavelike disturbances, large amplitude meanders and detached eddies

zonal flow instability is shown in Fig. 4.8. A similar study on the instability of the Azores current was performed with a HPE model by Kielmann and Käse (1987). Other investigations have been performed in a more complex environment (including topography and a curved coastline) for coastal current instabilities (Haidvogel et al. 1991b).

4.2.2.4 Isolated Eddy Translation

Once eddies are generated, they exhibit their own dynamics. Another class of models therefore examines the translation and dynamics of isolated eddies, neglecting the ambient mesoscale field. These models aim at an understanding of Gulf Stream Rings, Agulhas Rings, and Mediterranean Water eddies (Meddies), which are relatively long-lived entities that isolate and transport tracers.

Coherent vortices in the ocean have received much attention in the 1980s. Quasigeostrophic models of eddy translation and the interaction of individual vortices have been developed (McWilliams 1985; McWilliams and Gent 1986; Beckmann and Käse 1989). The westward movement of anticyclones can be explained by vorticity dynamics, adjustment processes during the translation, and the combined effects of interaction with mean flows, ambient mesoscale eddies, and an adjustment to the background stratification. For example, Fig. 4.9 schematically shows the translation of an anticyclonic eddy and the generation of Rossby waves to the east of the vortex.

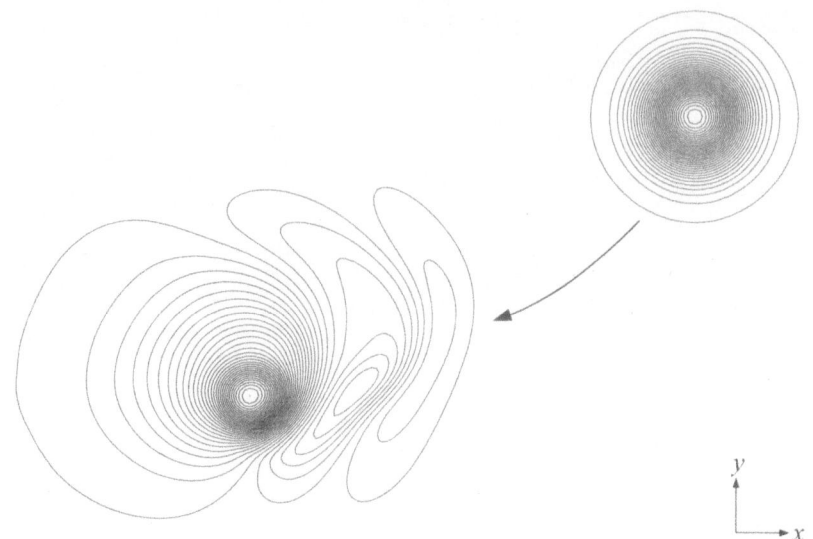

Fig. 4.9. Translation of an isolated baroclinic eddy on the β-plane, as seen in the density perturbation field. The anticyclonic vortex propagates westward and equatorward and generates a train
of Rossby waves. This behavior serves as a prototype for oceanic rings (Gulf Stream, Agulhas)
and submesoscale eddies (Meddies)

4.2.2.5 Surface Mixed Layer Dynamics

An area where one-dimensional process models are still of great importance is the
surface mixed layer of the ocean. Mixed layer models are of interest, because they
play a mayor role in physical ocean-atmosphere interaction, and in coupled physi-
cal-biological studies (see also Sect. 4.2.3.1).

A very simple model was presented in Sect. 4.2.1.2. It only considers the wind
input in a homogeneous ocean. For a stratified ocean, the buoyancy forcing is
similarly important. Two different approaches exist: vertically integrated Kraus-
Turner type (Kraus and Turner 1967) models and turbulent closure models (Mellor and Yamada 1982; Large et al. 1994). While the first assumes vertical homo-
geneity with the mixed layer, the latter does just the opposite and treats neutrally
stratified fluid as an exception. Although highly idealised due to the one-
dimensional (local) approach, these models are the basis for any treatment of the
surface mixed layer in the ocean.

The main issues that need to be addressed by mixed layer models are the annual
and diurnal cycle of the mixed layer (see Fig. 4.10). To answer these questions,
the models need to capture the net effects of small scale processes like small scale
turbulence, convective rolls, and double diffusion.

A milestone was the study by Woods and Barkmann (1986), who considered
seasonal changes in mixed layer quantities as the result of surface buoyancy and
momentum fluxes. A validation of these models is usually possible at certain loca-

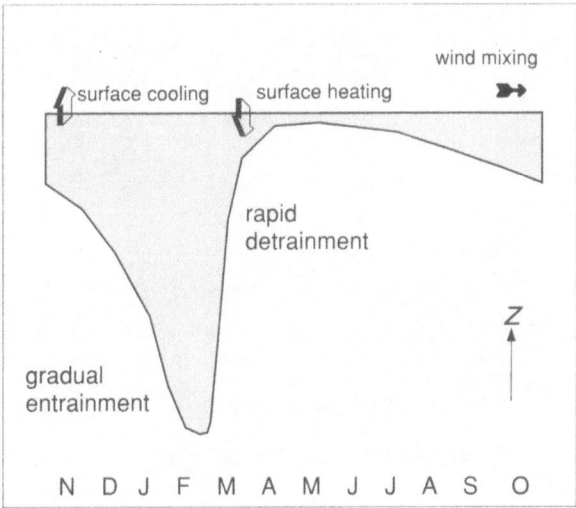

Fig. 4.10. Seasonal change of the surface mixed layer depth due to surface buoyancy and momentum fluxes

tions where lateral advection is weak. A recent overview on mixed layer models can be found in Large (1998).

4.2.2.6 Passage Throughflows and Sill Overflows

The global thermohaline circulation is critically controlled by flow through narrow passages and over shallow sills. To understand the processes that occur in connection with passage throughflows and sill overflows, idealised configurations have been used. Smith (1975) devised an integrated so-called "stream tube" model to simulate the behavior of outflow plumes. This model was one-dimensional; a single layer of homogeneous fluid. It could successfully describe the turning of such plumes due to the Coriolis acceleration, and could be tuned to give reasonable entrainment rates.

An extension to two-dimensional flows was done by Jungclaus and Backhaus (1994), who used an inverted reduced gravity model (see Appendix D) with one moving and one resting layer of fluid. It was applied to a "dam-break" problem, where a dense water mass on the shelf is released instantaneously. The resulting plume is schematically shown in Fig. 4.11.

Three-dimensionality is required to include the barotropic component of the flow which is spun-up. Recent efforts in this respect employ high resolution HPE models with topography-following vertical coordinates (Gawarkiewicz and Chapman 1995; Chapman and Gawarkiewicz 1995; Jiang and Garwood 1995; Jiang and Garwood 1996; Jungclaus and Mellor 1999.

The remaining goals for these process studies are to understand and quantify the entrainment mechanisms in such plumes. This is especially important as a correct parameterisation of down-slope flows, which are necessary for large-scale

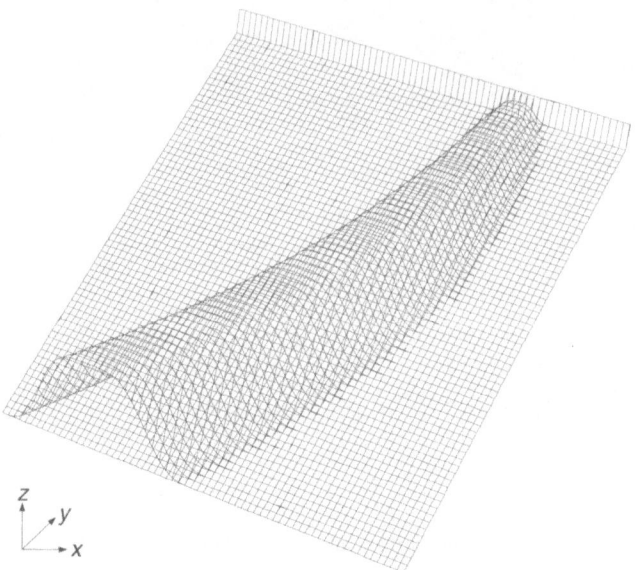

Fig. 4.11. Dense bottom water flows down topographic slopes, turns right (on the northern hemisphere) under the influence of the Earth's rotation, and entrains the ambient water masses, thus diluting the source water mass but increasing transport

climate models. See Beckmann (1998) for a review of the representation of such downslope flows in OCGMs.

4.2.2.7 Flow in the Vicinity of Steep topography

Steep topography in a rotating stratified ocean represents a special challenge for ocean models. Non-trivial analytical solutions are not known, and therefore a large variety of process models had to be developed to address this fundamental class of problems. Models exist for up- and downwelling processes at continental margins as well as for investigation of the effects of small scale features like coastal canyons and isolated seamounts.

Coastal up- and downwelling. Two-dimensional (x-z) models have been used extensively for this type of process study. The cross-shelf circulation is produced by a uniform wind blowing in the y-direction. An important result is the asymmetry of up- and downwelling circulations (see Fig. 4.12). The bottom boundary layer (BBL) is relatively thick in downwelling situations, as lighter water is transported below denser water and homogenised by vertical convection. At an upwelling coast, the bottom boundary layer is comparatively thin, because the upwelling denser water increases the stability on the shelf and slope, (Trowbridge and Lentz 1991; Ramsden 1995a,b).

For alternative forcing, additional effects will emerge. Even in purely periodic wind, the asymmetry of upwelling and downwelling circulations as described may

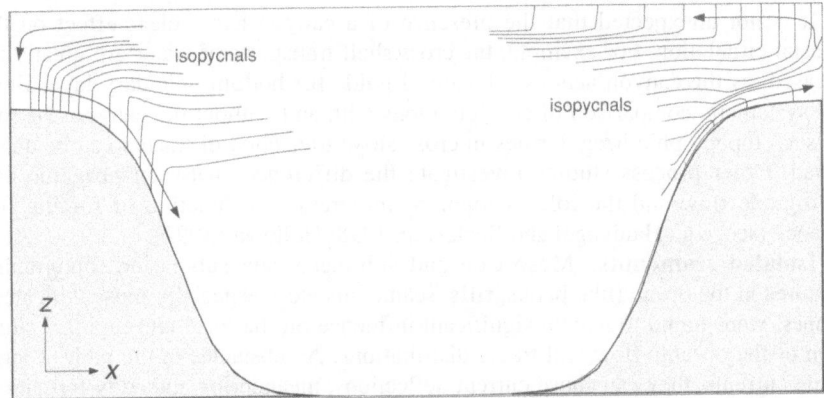

Fig. 4.12. Overturning circulation in the coastal ocean. Along-shore wind causes upwelling at left-hand coasts and downwelling at right-hand coasts. The bottom boundary layer differs drastically.

lead to a substantial net circulation. This residual flow is well known to exist in areas with large tidal amplitudes, but any periodic or stochastic forcing will generate such a rectified flow, especially for steep and irregular topography.

Submarine canyons. The influence of small-scale canyons in the continental margins has been explored in detail by Klinck (1996). In this case, the above two-dimensional model needs to be extended into the along-coast direction. Often, a periodic channel configuration on an f-plane is chosen (see, e.g., Fig. 4.13a).

Typically, an initial stratification without lateral gradients is prescribed. The model is either run into steady state (for constant forcing) or has to be averaged over a sufficiently long period to obtain the meaningful time-mean fields (for variable forcing). Processes directly related to steep topography can best be studied ied with topography-following coordinate models (see Haidvogel and Beckmann 1998).

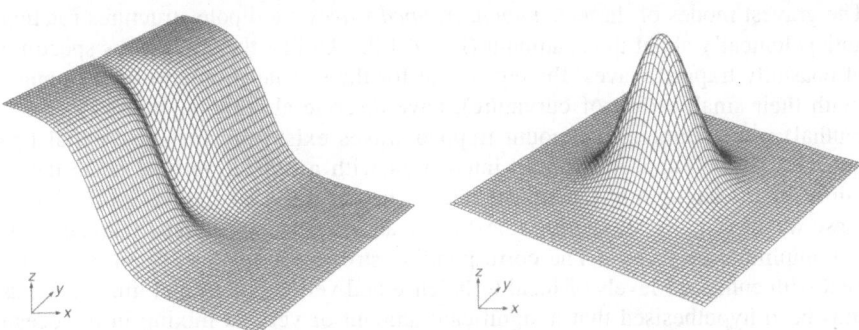

Fig. 4.13. Prototype topographies for the study of effects of steep topography on oceanic flows. (a) a continental shelf/slope configuration with a coastal canyon and (b) an isolated submarine seamount

It is not unexpected that the presence of a canyon has a clear effect on the coastal circulation. For example, the cross shelf transports of water are greatly increased, as the canyon acts as a localised guide for bottom boundary layer flow. By systematic comparison of configurations with and without the canyon, the role of such topographic irregularities in cross-slope transports of tracers can be quantified. Other process studies investigate the differences between prograde and retrograde flow and the role of topographic stress as a function of forcing frequency (see, e.g., Haidvogel and Beckmann 1998; Holloway 1992).

Isolated seamounts. Mesoscale and sub-mesoscale submarine topographic features in the ocean (like banks, riffs, seamounts etc.), especially those with steep slopes, were found to have a significant influence on the local and regional situation of the oceanic flow and tracer distributions. As obstacles in the path of large scale currents, they can cause current deflections, meandering and eddy formation; thus increasing the variability in the vicinity and downstream of the topographic irregularity. On the other hand, many of them have their own isolated micro-environment, quite distinct from the nearby surroundings (Roden 1987). It is thus not surprising that they are important for both physical and biogeochemical sciences.

Theoretical models are available for both steady and periodically forced flow at topographic obstacles. The most basic fluid dynamics principle directly relevant to isolated topographic features dates back to Taylor (1917). He showed that steady, linear and inviscid flow has to follow isobaths; areas of smaller (*larger*) depth exhibit anticyclonic (*cyclonic*) circulation. In case of closed depth contours, fluid parcels above these areas have to remain within a so-called *Taylor column* (Fig. 4.14a). Even though unsteadiness, non-linearity and viscous processes in the ocean will relax this constraint, the underlying physical mechanism remains important and can be observed in geophysical fluids. In a generalisation of Taylor's theory, trapped water bodies of arbitrary shape above seamounts in stratified fluids are called *Taylor caps* (see Schär and Davies 1988).

Another theoretical approach considers the propagation of trapped waves about isolated topographic structures. Like coastally trapped waves (see LeBlond and Mysak 1978) at the continental shelf, oscillatory solutions of the linearised equation of motions in a rotating fluid exist at isolated topography (Brink 1989; 1990). The gravest modes of these *seamount trapped waves* are dipole-structures rotating anticyclonically about the seamount (Fig. 4.14b). Unlike the continuous spectrum of coastally trapped waves, the equivalent for the special geometry of seamounts (with their small radius of curvature), have discrete along-topography (i.e., azimuthal) wavenumbers. Seamount trapped waves exist only for sub-inertial frequencies $(\omega < f)$ and are bottom intensified with a typical vertical scale determined by the strength of the stratification. Theoretically, a continual excitation of these waves (e.g. by alternating tides) can lead to resonance effects, with substantial amplification factors. The corresponding strong currents can often be associated with enhanced levels of local turbulence and vertical/diapycnal mixing. It has thus been hypothesised that a significant amount of vertical mixing in the ocean occurs at seamounts. If effects of non-linearity are taken into account, time-mean effects of these waves are possible. Several theoretical studies on this subject (Wright and Loder 1985; Maas and Zimmermann 1989a,b) have led to the idea that rectified currents (see Fig. 4.14a) at seamounts may be substantial.

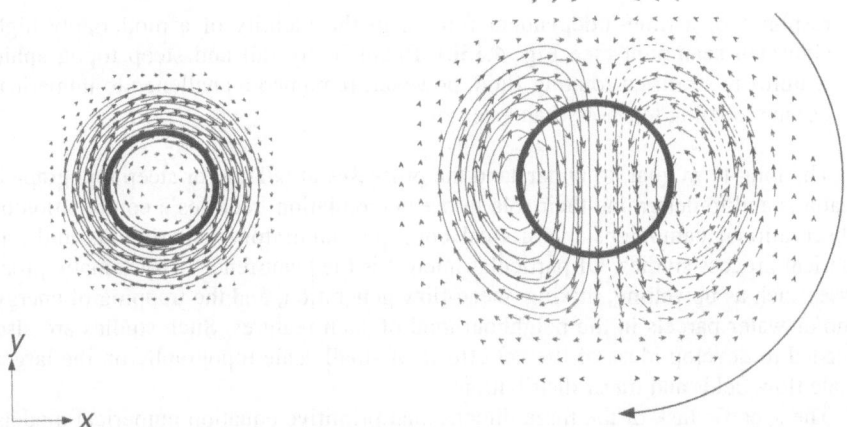

Fig. 4.14. Typical flow fields at an isolated circular seamount: (a) stationary vortex (Taylor column, Taylor cap in stratified fluids), which can develop both as the result of steady impinging flow or through the rectification of periodic seamount trapped waves. The rotational sense is anticyclonic on the northern hemisphere; (b) first azimuthal mode seamount trapped wave (a special form of a coastal trapped wave)

Based on these theoretical ideas, three different aspects of flow around isolated seamounts are of interest: the steady state (time-mean) circulation, the transient response to time-independent forcing, and the wave generation for periodic forcing. Investigations in all three areas have been performed, with a variety of approaches.

- Analytical solutions of flow around seamounts exist for simplified dynamical systems. Fennel and Schmidt (1991) used quasigeostrophic theory to investigate the transient response of uniform flow, impinging on a right circular cylinder. Their results show the occurrence of a dipole structure of vertical velocity and density perturbation propagating anticyclonically about the obstacle; until it is locked in place, the positive lobe moves over the flat region of the fluid and a Taylor column is formed. A steady state layered (reduced gravity) approach was taken by Ou (1991), who could relax the quasigeostrophic limitation of small amplitude topographic variations and obtain steady state solutions of Taylor caps.
- Numerical quasigeostrophic simulations were carried out by McCartney (1975). It was shown that isolated topographic features in a strong flow can produce down-stream wave generation and may thus act as triggers for observed meandering of major ocean currents.
- Laboratory experiments have concentrated on periodically forced flow about seamounts (Boyer and Zhang 1990; Codiga 1993) and have qualitatively confirmed the theoretical ideas on trapped wave generation and mean flow rectification.
- A first series of numerical experiments with an HPE model (see Appendix) were performed by Huppert and Bryan (1976). They considered the transient

response of a time-independent forcing in the vicinity of a moderately high Gaussian seamount (see Fig. 4.13b). Realistically tall and steep topographic features in stratified rotating fluid, however, remained a challenge to numerical ocean modeling for more than a decade.

The long-term goal, to understand the processes at tall and/or steep topographic features and to determine their role in ocean circulation, requires a combination of observational campaigns, theoretical concepts, laboratory experiments and numerical process studies. Of particular interest is the occurrence of systematic processes such as upwelling, mixing, mean flow generation, and the trapping of energy and/or water parcels in the neighbourhood of such features. Such studies are also needed to develop ideas of the net effects of small-scale topography on the larger scale flow fields and tracer distributions.

The specific task of the three-dimensional primitive equation numerical models is to extend the theoretical investigations beyond their intrinsic limitations (weakly non-linear, weakly stratified, small amplitude topography, quasigeostrophic).

Systematic studies on the circulation around tall and steep seamounts in a rotating (f-plane), stratified fluid have been performed for idealised geometries (Chapman and Haidvogel 1992; 1993; Haidvogel et al. 1993; Beckmann 1995), both for steady and periodic forcing (i.e., the generation of seamount trapped waves and/or Taylor caps). The influence of both physical and numerical aspects has been investigated in detail. A recent study (Beckmann and Haidvogel 1997) was even successful in quantitatively explaining the observed flow (Brink 1995; Kunze and Toole 1997) around a seamount in the Northeast Pacific, *Fieberling Guyot*, including the amplitude of the trapped waves, the strength of the rectified mean flow and the occurrence of higher harmonics in temperature recordings atop the seamount. This was possible, because tidally forced flow is the dominant mechanism at the location of this seamount in the northeast Pacific Ocean.

4.2.2.8 Open Ocean Convection

Static instability of the water column occurs as the result of surface thermohaline forcing (cooling, evaporation, brine release from sea ice) or lateral advection processes, which place denser above lighter water. Such unstable situations do not persist for a long time; they are removed by vigorous and small-scale vertical motion, called "convective plumes," which transport the denser water downward, leading to a vertical homogenisation of the water column (see Fig. 4.15 for a schematic representation of this process). The compensating upwelling motion does not occur localised but as a large scale and weaker upward flow. The penetration depth of the convection depends on the initial stratification of the fluid ("preconditioning") and the strength and time scale of the cooling/evaporation (Schott et al. 1994; Send and Marshall 1995). For an overview, the reader is referred to Send and Käse (1998).

To study these processes, the full NHPE system needs to be utilised and analytical approaches (aside from scaling considerations) are not possible. Non-hydrostatic models, in which the total time derivative of the vertical velocity is re-

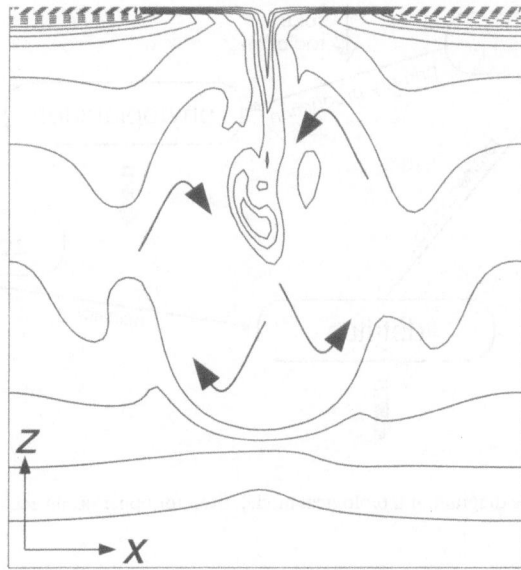

Fig. 4.15. Deep convection due to surface cooling occurs through vigorous downwelling plumes on horizontal scales of a few tens of meters and leads to a homogenisation of the water column

tained, have shown that highly nonlinear dynamics governs three-dimensional rotating convection in geophysical fluids (Jones and Marshall 1993; Sander *et al.* 1995).

It is interesting to note that these studies have led to the conclusion that convective plumes can be parameterised successfully with a large (1 to 100 $m^2\ s^{-1}$) vertical mixing coefficient within the usual harmonic mixing concept (Send and Käse 1998).

4.2.3 Interdisciplinary Process Models

Marine research is not limited to physical processes. Aspects of biological, geological and sea-ice research are closely linked to ocean physics and can benefit from a combination with the process-oriented models described in the preceding chapter.

4.2.3.1 Biological Models

Aspects of marine biology have recently been studied within the framework of process-oriented models. These are based on the nitrate or carbon cycle and involve a number of compartments (see Fig. 4.16 for example) as well as interactions between these. Expressed in a set of differential equations, the time evolution of the system is based on just one single conservation principle.

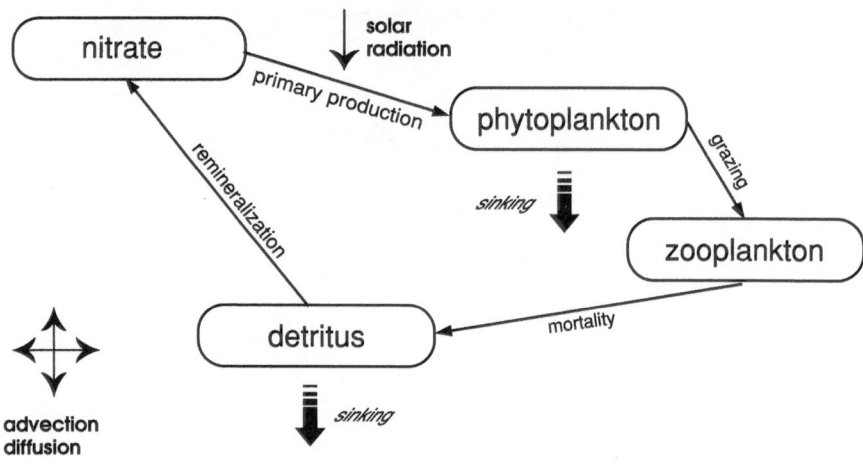

Fig. 4.16. Schematic diagram of a biological model, used for upper ocean ecosystem studies

Zero- or one-dimensional (vertical) configurations (see also Sect. 4.2.2.5) have been used extensively. The focus is then: dynamics of phytoplankton bloom (phase, amplitude, dependencies on nutrient fluxes, etc.). Physical processes are included in one-dimensional models in the form of vertical mixing and (occasionally) vertical advection terms.

These studies are complicated by the fact that often large uncertainties exist in the values of transfer coefficients and the necessary number of compartments.

4.2.3.2 Sediment Transport Models

The erosion, transport, and deposition of sediment is one of the main interests of marine geology. Consequently, coupled physical-geological studies have been driven by geological questions. There is, however, an important process relevant for ocean dynamics: sediment plumes (turbidity currents) may contribute significantly to the deep water formation in high latitudes (Fig. 4.17).

Near-bottom down-slope flows can reach substantial velocities at the bottom, depending on the density contrast relative to the ambient water masses, the steepness of the topographic slope, and the roughness of the bottom. When the velocity at the bottom of the ocean exceeds a critical value, erosion of sediment begins. This critical velocity depends on the grain size of the sediment, but is typically between 10 and 20 cm s^{-1} for grain sizes between 10 and 100 μm. The additional suspended material increases the density of the fluid; Fohrmann et al. (1998) cite values of more than 1 kg m^{-3} for medium density suspension flows. Converted to salinity equivalents, the sediment load might add up to 1 psu to the density of the moving, turbulent BBL. Thus it seems plausible that these plumes may gain additional speed from an even larger density contrast to the ambient field, leading to

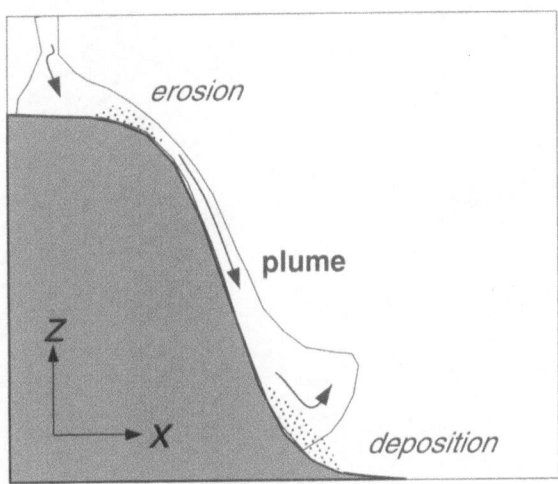

Fig. 4.17. Schematic representation of the processes within a sediment plume that transports particles from the shallow shelves down to greater depths

deeper penetration into the abyss. Sediment-laden bottom plumes are thus potentially important for setting the deep water characteristics; a decelerating plume will lose its sediment load and *upward convection* may occur.

It is assumed that sediment transport can be modeled by an advection-diffusion equation with special source and sink terms:

$$\frac{\partial}{\partial t}(hC) + \frac{\partial}{\partial x}(u_b hC) + \frac{\partial}{\partial y}(v_b hC) = A\nabla^2(hC) + \text{sources} - \text{sinks}, \tag{4.8}$$

where h is the thickness of the BBL and u_b, v_b are the (divergent) BBL velocities. The source terms are based on critical velocities for erosion (specified individually for each sediment class); the sink terms are modeled similarly by prescribing settling velocities and deposition probabilities. Fohrmann et al. (1998) have added such a sediment transport model to the reduced gravity model of Jungclaus and Backhaus (1994), and find substantial effects on the down-slope penetration of individual plumes. It has been speculated that sediment plumes are important for determining the properties of deep water masses, and for the supply of the deep water with oxygen and other nutrients.

More work is needed on the effects of different grain sizes, vertical distribution within sediment plumes, the influence of topographic irregularities, and the occurrence of internal nepheloid layers.

4.2.4   Sea Ice Models

Another class of processes relevant for physical oceanography involves the dynamics and thermodynamics of sea ice. Dynamic-thermodynamic sea ice models need to capture (the net effects of) the following processes/phenomena: ice growth

Fig. 4.18. Sea ice floe field, with floes of differing size and thickness (indicated by greyscale)

and decay (ice thickness and compactness, temperature and salinity distribution within the ice); ice drift and dynamics (ice roughness, formation of pressure ridges and leads; development and maintenance of polynyas and landfast ice), melt water pools, as well as the snow cover, and the snow conversion to ice (flooding).

It is therefore surprising that almost no systematic process studies on sea ice dynamics and thermodynamics exist. Most currently used sea ice models are applied in regional and coupled studies. They are based on continuous equations for ice thickness, ice compactness and ice velocities, as they assume fields of rather than individual ice floes (see Fig. 4.18). This is despite the fact that the continuum hypothesis is only justifiable for large scale models of this granular material. Lagrangian methods, which seem to be more appropriate for description of an ensemble of ice floes, are not as well developed as the continuum mechanics.

4.3 Concluding Remarks

4.3.1 Outlook

The range of process-oriented models spans the scales and complexity from simple linear waves to fully three-dimensional and nonlinear configurations. Despite (or because of) the involved idealisations, these models have been found extremely useful to gain understanding of the dynamics of certain aspects of ocean dynamics, and to assist in the interpretation of (sparse) observational data.

The limitations of analytical tools are partially eliminated by numerical process models, which can be used to

- extend analytical studies into the nonlinear regime;
- explore the parameter dependence of complex mechanisms; and
- test and help in the development of parameterisations.

A generalisation of the results of a process-oriented model is usually desired; however, it is often complicated due to the necessary simplifications of the configuration. A (quantitative) validation of results is therefore only possible if circumstances are favourable (i.e., if a region exists where one single process is dominant).

The continuing development of process models will lead to a more and more complete understanding of individual phenomena and the combined effects within multi-process and multi-component marine systems.

4.3.2 Comments on Numerical Process Models

The large number of these process studies that are currently performed with numerical models necessitates a few comments on the dependence of model results on numerical choices and parameters. A numerical approach offers in principle an unlimited flexibility with respect to the inclusion of nonlinearity of the flow, complex topography and geometry, and highly variable (in space and time) forcing. There are, however, aspects of the numerical model which might influence the representation of physical processes. These are:

- resolution (mainly spatial). A scale selection is implicitly made by choosing a resolution. The assumption is that the sub-grid scale processes are either not relevant or can be successfully parameterised.
- horizontal grids. A regular rectangular grid leads to a step-like representation of boundaries (see Fig. 4.7), which can cause systematic errors. Boundary fitting (conforming) grids are available and may be better suited for problems that are sensitive to coastal boundary conditions. The multi-scale approach of finite element grids represents the most flexible approach in this respect.
- vertical coordinate systems: geopotential levels (this is the most simple vertical coordinate, which has been applied to a large number of process studies; there are, however, severe limitations with respect to the representation of topography, as the resulting step-like approximation to topographic slopes is clearly suboptimal.)
- terrain-following (these are best suited for studies of topographic waves near bottom flows, as the lower boundary is a coordinate surface.)
- potential density layers (are most useful for investigations of upper ocean frontal systems and subduction processes, as spurious diapycnal mixing can be completely eliminated.)

- numerical algorithms. E.g., advection schemes, which ideally should be monotonic and nondiffusive at the same time. However, such a scheme does not exist and compromises have to be made which can influence the numerical solution.
- subgridscale (sgs) parameterisations. The main goal of sgs parameterisations is to represent oceanic mixing processes in an adequate way. Mixing in numerical ocean models, however, is required for both physical and numerical reasons. Physically, subgridscale processes often lead to net mixing. Numerically, grid scale noise needs to be damped to ensure a stable integration. Adaptive methods, which depend on the local flow field and tracer distributions, seem most appropriate for this purpose. Finally, in addition to these turbulence parameterisations, non-diffusive processes like convection also need to be parameterised.

Parameter studies that concentrate on numerical aspects of a process model are often labeled as "purely technical." They are, however, extremely important for today's global climate or coupled physical-biogeochemical models, which cannot be run at eddy resolving resolution. The consequences of missing eddies, diffused fronts, closed or widened passages need to be investigated in systematic process studies.

Acknowledgments

I thank Hans von Storch and Götz Flöser for organising and inviting me to this truly multidisciplinary summer school. Helpful comments by Ralph Timmermann on an earlier version of this manuscript are gratefully acknowledged.

Appendix

A   HPE

Applying the Boussinesq, thin shell, hydrostatic and traditional and incompressibility approximations leads to today's most widely used system of equations for OGCMs. The Earth's shape is approximated by a perfect sphere, so that gravity points to the center of mass. Also, tidal motions are usually neglected. The NHPE system (1-5) then becomes:

$$\frac{du}{dt} - fv = \frac{1}{\rho_0} p_x + F^u + D^u \qquad (4.9.1)$$

$$\frac{dv}{dt} + fu = \frac{1}{\rho_0} p_y + F^v + D^v \qquad (4.9.2)$$

$$0 = \frac{1}{\rho_0} p_z - \left(\frac{g}{\rho_0}\right)\rho \qquad (4.9.3)$$

$$\nabla \cdot \vec{v} = 0 \qquad (4.10)$$

$$\frac{dT}{dt} = F^\tau + D^\tau \tag{4.11}$$

$$\frac{dS}{dt} = F^S + D^S \tag{4.12}$$

$$\rho = \rho(S, T, p) \tag{4.13}$$

These so-called *primitive equations* are well-suited for all process models except for convection studies.

B   QG

Further approximation (first order in Rossby number) and manipulation of the equations (4.9-4.13) yields the quasigeostrophic system: Taking the curl of the momentum equations and defining a *streamfunction*:

$$\psi = \frac{p}{\rho_0 f_0} \tag{4.14}$$

Such that

$$(u, v) = (-\psi_y, \psi_z) \tag{4.15}$$

and

$$\varsigma = \nabla^2 \psi, \tag{4.16}$$

we arrive at the quasigeostrophic potential vorticity equation. (see e.g. LeBlond and Mysak 1978):

$$\frac{d}{dt} \left\{ \nabla^2 \psi + f + \frac{\partial}{\partial z} \left[\frac{f_0^2}{N_0^2} \frac{\partial \psi}{\partial z} \right] \right\} = 0 \tag{4.17}$$

In this system, potential density alone replaces the thermodynamic quantities T and S (single state variable) and becomes a diagnostic quantity:

$$\rho' = -\frac{f\rho_0}{g} \frac{\partial \psi}{\partial z} \tag{4.18}$$

Separating horizontal and vertical dependence (for flat bottom) results in an eigenvalue problem for the vertical structure (see Fig. 4.2b):

$$\frac{\partial}{\partial z}\left[\frac{f_o^2}{N_o^2}\frac{\partial F}{\partial z}\right] + \lambda^2 F = 0.$$

The solutions, corresponding to different eigenvalues λ are so-called *vertical modes*. The inverse eigenvalues are called *Rossby radii of deformation* and set the typical horizontal scale for each vertical mode:

$$R_D \equiv \lambda^{-1} = \frac{\sqrt{gh_e}}{f},$$

where h_e is the *equivalent depth* for stratified fluids (see e.g. LeBlond and Mysak 1978).

C   SWE

The so-called *barotropic vorticity equation* is derived from the hydrostatic primitive equations by vertical integration and further assumptions relating to the vertical structure of solutions. The resulting equation is two-dimensional (x,y), yet retains a lot of the dynamical complexity of three-dimensional flow on the rotating earth. Defining

$$\vec{V} = \int_{-H}^{\varsigma} \vec{v}\, dz \tag{4.19}$$

the resulting depth-averaged horizontal momentum equations are

$$\frac{\partial}{\partial t}(H\vec{V}) + \nabla \cdot (H\vec{V}\vec{V}) + fH\hat{k} \times \vec{V} + gH\nabla\eta - \vec{\tau}^w + \vec{\tau}^b = 0 \tag{4.20}$$

where a vertical viscous term was included to explicitly incorporate vertical momentum exchanges (stresses) at the surface ($\vec{\tau}^w$) and bottom ($\vec{\tau}^b$).

Eliminating surface gravity waves by applying a rigid lid upper boundary condition, motions become horizontally non-divergent, and a streamfunction (see Eq. 4.14) can be introduced, which leads to the barotropic form of Eq. (4.17).

D   RG

The reduced gravity model is obtained by the assumption of one moving layer of homogeneous fluid lying above (or *below*) an infinitely deep layer of denser (*lighter*), but resting fluid. In this case, the gravitational acceleration is replace by the reduced gravity

$$g' = g\frac{\delta\rho}{\rho_0}. \tag{4.21}$$

Chapter 5
Mathematical Models in Environmental Research

by H. Langenberg

5.1 Introduction

Mathematical models are widely used in environmental research, and most of the lectures in this school are concerned with models that are in some way based on mathematics. In this contribution, the idea of mathematical modeling will be presented, along with a variety of examples that use different fields in mathematics as the basic means of representing reality.

This rather general (but surely not complete) overview is then complemented with a more detailed analysis of one particular kind of mathematical model, namely a numerical model of the ocean that is based on the primitive equations. Here, two steps are distinguished: first, the description of nature in differential equations (associated with the mathematical field of Dynamical Systems) and second, their numerical realisation to suit a computer and make their solution feasible (associated with Numerics).

5.2 Mathematical Models – an Overview

According to the Oxford Advanced Learner's Encyclopedic Dictionary, a model (in our context) is "... a fragment of a mathematical or formal theory that reflects some aspect of a particular physical, social, technological or natural phenomenon. [...]"

This definition reveals two important features of models: it is usually only "a fragment" of a theory that is used to build a model, and the model only represents "some aspect" of a phenomenon. This restriction that is inherent in the simplifying nature of models is often forgotten, and the results are interpreted as if they were observations. It is, however, essential to distinguish between those effects that are included in the model and those that are clearly external to its setup. Only with this distinction is a meaningful interpretation of the results possible.

In the following, several types of mathematical models and their applications are shortly introduced.

5.2.1 Box Models

The least complex type of model that is referred to here is the so called "Box Model." Such a model consists of a number of compartments of the environment that are individually addressed (and can therefore not be very numerous: typically three to ten boxes are used), together with a mathematical formulation of their interactions. Each compartment evolves in time from a set of initial conditions (obtained from observations) via fluxes to and from the other compartments.

One example of a box model is presented in Fig. 5.1 (from Schulz 1998, pers. comm.): a budget of lead in the North Sea is calculated from emissions, atmospheric deposition, exchange with surrounding seas and rivers and deposition in the sediment. The interactions between the boxes "atmosphere," "North Sea," "sediment" and "Atlantic Ocean" are calculated according to rather simple mathematical formulae and depend essentially on the concentrations in the various compartments. With a time step of one year, it is then possible to reconstruct a simulated sediment core for the North Sea that reacts to the development of lead emissions.

This type of budget model gives a rough idea of the dynamics of a quite restricted environmental system (in this example: lead distributions in the North Sea region) and puts various observed details, as for instance the deposition into the sea relative to the atmospheric lead concentration, into the context of the system. However, many aspects that contribute have to be omitted in this type of model, and each compartment is only allowed one "box" in the model.

5.2.2 Cellular Automata

Cellular automata represent a quite different mathematical framework for modeling the natural or social environment. They consist of a number of cells and a set of values that the cells can have (in the simplest case just "0" and "1"). For each cell i, a finite "neighbourhood" $N(i)$ is defined, which contains all those cells that have an influence on the cell i within one time step. To formulate the time dependence, a local transition function f_i is defined that calculates the value of the cell i at the next time step according to the values of the cells in the neighbourhood $N(i)$ at present. All the cells are then changed simultaneously, each according to its local transition function.

This setup allows for a much larger number of cells than boxes in a box model, while retaining some individuality of the cells: it is in principle possible within the framework of cellular automata to give each cell a different neighbourhood and a different transition function. However, most common are cellular automata with uniform neighbourhoods (e.g. for an automaton that is defined on integers, $N(i)$ could be the set $i - 1$, $i + 1$ and uniform local transition functions (i.e. the sum modulo 2 for the above mentioned automaton).

An overview and many types of cellular automata for the various disciplines are introduced for instance in Casti (1989). One of his examples is explained in the following. More applied studies using cellular automata that are more focussed on environmental system modeling are given by Engelen et al., (1995, 1997).

Here, a cellular automaton that simulates the generation of vertebrate skin patterns is introduced in order to illustrate how cellular automata work. The (much

Lead in the North Sea: A tentative balance

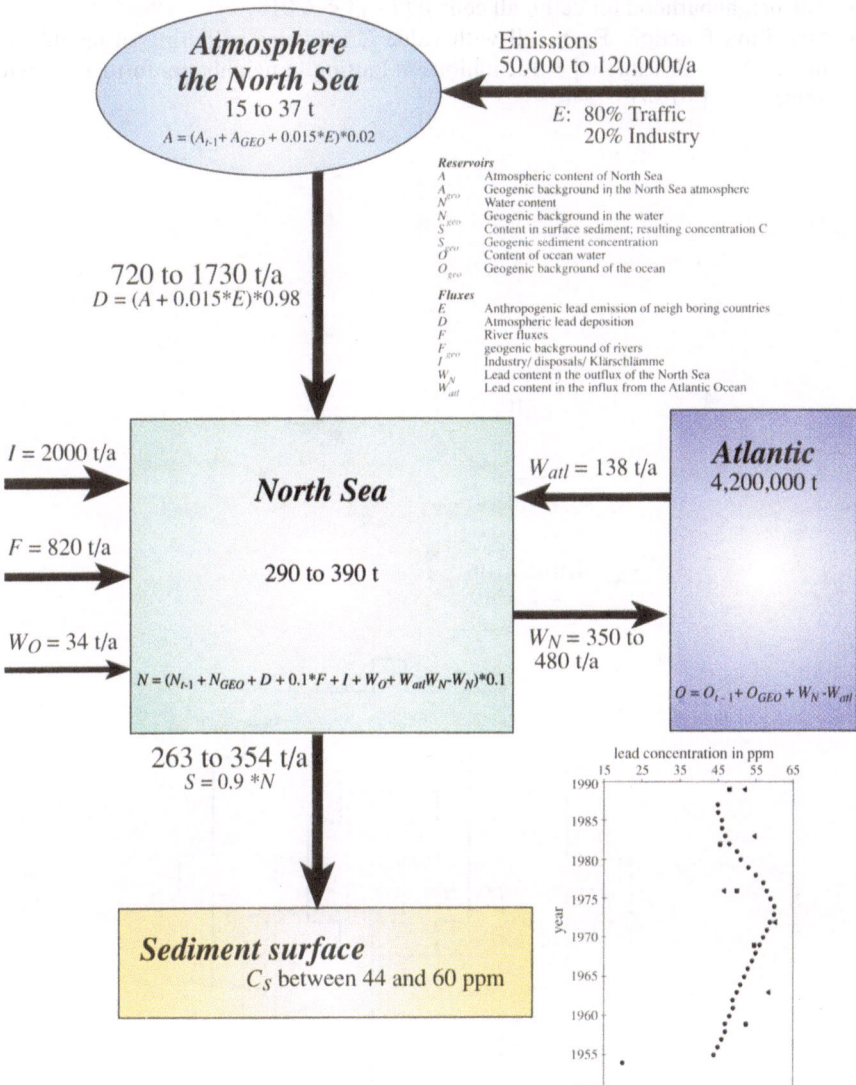

Fig. 5.1. Lead content in the North Sea region: an attempt at a budget

simplified) biological idea of what is happening can be explained as follows. In the vertebrate embryo, pigment cells are distributed randomly over the skin. Each pigment cell sends out two types of morphogenes: a colour inhibiting one, with a relatively large radius of action; and a colour activating one, whose influence does not extend as far, but that has a larger impact (see Fig. 5.2, top). The cellular automaton to simulate this behavior is defined as follows:

- The cells: a finite sector of IN x IN
- The values: C (coloured) and U (uncoloured)
- The neighbourhood for cell i: all cells j: $|i - j| < 6.01$
- Transition function: Each cell with value C activates colouring in neighbours nearer 2.3 with value $w_1 = +1$, inhibits colouring in neighbours further 2.3 with value $w_2 \in \{-0.34\}$.

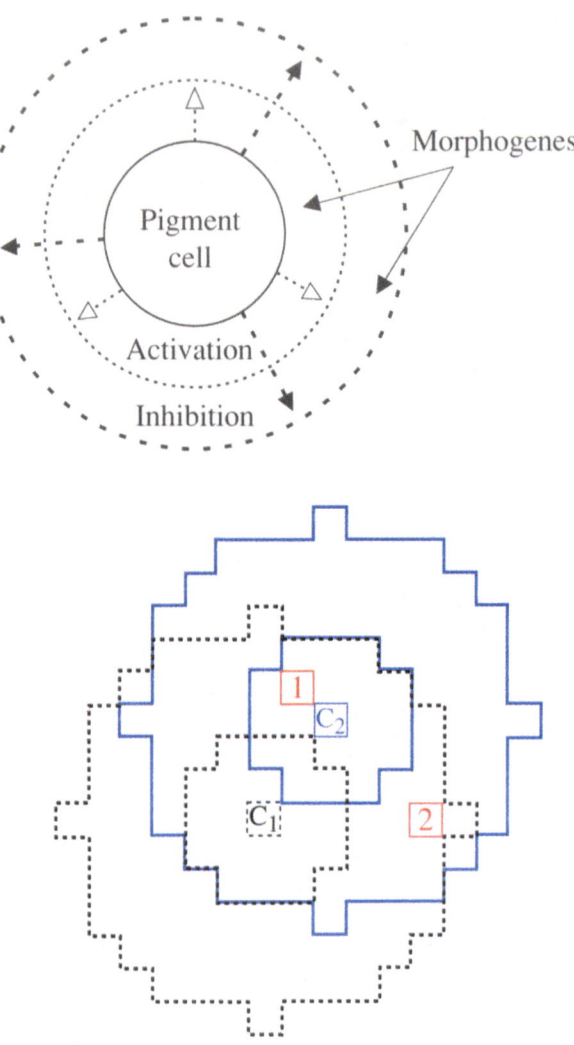

Fig. 5.2. Top: A pigment cell with two circles of influence: colour activation in the closer surroundings and a colour inhibition in the area further away from the cell. Bottom: The associated cellular automaton: c_1 and c_2 are pigment cells. Then cell 1 receives activation $w_1 = 1$ from c_2 and inhibition $w_2 = -0.34$ from c_2, thus the value of cell 1 at $t = 1$ is: $v(1)_{t+1} = 1 - 0.34 > 0$ and cell 1 will be coloured in the next time step. Cell 2 receives inhibiting signals from both c_1 and c_2, giving $v(2)_{t+1} = w_2 (c_1) + w_2 (c_2) = -0.34 - 0.34 < 0$ and no colour at $t+1$

The value of i at time $t+1$ is calculated as:

$$v(i)_{t+1} = \sum_{j \in N(i)} w$$

If $v(i)_{t+1} > 0$, then i at $t+1$ is coloured,
If $v(i)_{t+1} = 0$, then i does not change.
If $v(i)_{t+1} < 0$, then i at $t+1$ is uncoloured.

In Fig. 5.2, bottom, the evolution of a pattern is demonstrated for an initial condition with only two coloured cells: Cell 1 receives an inhibiting signal with value $w_2 < 0$ from one coloured cell ($C1$) and an activating signal with value $w_1 > 0$ from the other one ($C2$). As $/w_1/ > /w_2/$ by definition, the added effect on cell 1 is activating, so that it will be coloured in the next time step. Cell 2, on the other hand, receives only inhibiting signals and is thus left uncoloured. It is easy to see that this automaton favours the conglomeration of coloured cells. The resulting patterns for four different values of w_2 are shown in Fig. 5.3.

This automaton does not simulate a true evolution in time, but is just left to adjust until a quasi-stationary state is reached. It is, however, possible to simulate a development by introducing suitable boundary conditions that insert new information into the system at each time step. Otherwise, a finite cellular automaton always ends in a steady state or a cycle of finite length (Langenberg 1996).

$w_2 = -0.34$ -0.28 -0.24 -0.20

Fig. 5.3. Skin patterns emerging from different inhibitor values (from Casti 1989)

5.2.3 Differential Equations

Probably the most widely used technique for mathematical simulation of the physical environment is the use of differential equations that describe the development in time of some entity, according to fixed laws. Since the rapid progress of electronic computing makes it possible to solve differential equations numerically, this way of simulating the world has been particularly successful and is applied in virtually all physics-based subjects from climate research to engineering.

A differential equation based ocean circulation model, with its generation from the ideas we have about nature to its numerical realisation, is treated in detail in Sect. 5.3. Naturally, models of different entities are different, but most features of modeling with differential equations are common to all such models.

5.2.4 Other Techniques

Other quite different techniques of mathematical modeling include statistical modeling and graphical modeling (which is based on statistical modeling and provides a way to present the results of statistical analyses more concisely). In fact, most of the lectures in this school deal with more or less mathematical descriptions of the environment and thus, with mathematical models.

The most important common properties of these models are:

- They reduce the environment to a finite number of components.
- They give a mathematical description of part of the dynamics of the system.
- They usually describe the development in time, thus in principle allowing predictions.

5.3 From Nature to Navier-Stokes' Equations

When trying to build a mathematical, differential equations-based model of the ocean circulation, the first question to ask is what is there that might influence the motion in the sea. A somewhat arbitrary list of processes and things that spring to mind is: ocean currents, turbulence, waves, wind, the bottom of the sea, land (beach), human impact (e.g.: dredging, fishing), tectonics, river runoff, solar heating, the Coriolis force, fish and plants, buildings (e.g. oil rigs, ships), sea ice, and sea surface elevation.

The first, the ocean currents, are considered to be what we want to simulate, so this is what the equations are to be solved for. The other items are only important if they contribute "significantly," and on the time scales we are interested in, to the ocean currents. Typically for a shelf sea model, time scales on the order of years or decades are simulated, so that tectonics is too slow to have an impact. On the other end of the scale, turbulence is thought of as everything whose temporal and spatial scales are too small to be treated explicitly, but it has to be taken into account in some way because the small scales feed back into the larger scales. This is usually done with a "parameterisation," i.e. trying to approximate the small scales in terms of the (known) large scale entities.

In the type of model that is introduced here, i.e. a primitive equation model of the North Sea, typically waves, human impact, fish and plants, sea ice and buildings are also neglected, because their influence on the currents is considered to be small. The other components that are mentioned above are expressed as terms in a differential equation. How they are generated will be explained in the following.

It is not a small problem in modeling an environmental system to decide which effects have to be taken into account and which can be left out. The environment is an open system that consists of infinitely many feedbacks with varying importance and it is very easy to miss a process that matters. However, when a model produces realistic results, it is generally assumed to include everything that would make a big difference.

One difficulty, for instance in climate change experiments, is that an effect that is negligible today may become important under the conditions that prevail in a different climate. A model that produces realistic results for present day conditions may thus be unsuitable for the simulation of a climate change scenario.

5.3.1   The Ocean Currents

The ocean currents are the parameter we want to simulate. Precisely, we want to simulate how they change in time under the influence of a given (and time-varying) set of boundary conditions. For this, the velocities are split into three components that are orthogonal to each other, resulting in the vector $\mathbf{V}=(u,v,w)$, where u, v and w are the zonal (x), meridional (y) and vertical (z) components, respectively, and positive u is assumed to point east, positive v north and positive w upwards.

The change in time of $\mathbf{V}$ can then be expressed as $d\mathbf{V}/dt$, the total derivative with respect to time. Expressed as partial derivatives and split up into the three velocity components, this looks slightly more complicated:

$$du/dt = \partial u/\partial t + u\partial u/\partial x + v\partial u/\partial y + w\partial u/\partial z$$

$$dv/dt = \partial v/\partial t + u\partial v/\partial x + v\partial v/\partial y + w\partial v/\partial z$$

$$dw/dt = \partial w/\partial t + u\partial w/\partial x + v\partial w/\partial y + w\partial w/\partial z$$

The total derivative consists of the local change in time, and the advective terms: the velocity field moves with the currents similarly to all other entities. Figure 5.4 illustrates how this works for the term $v\partial u/\partial y$: The field of u-velocities is moved in y-direction by the v-velocity.

The advective terms make the equations of motion non-linear, rendering them impossible to solve analytically in full complexity. This also means that the small scales (usually described as turbulence) influence the large scales. If the equations are solved numerically, turbulence will have to be parameterised as mentioned above. One way to do this is explained in the next subsection.

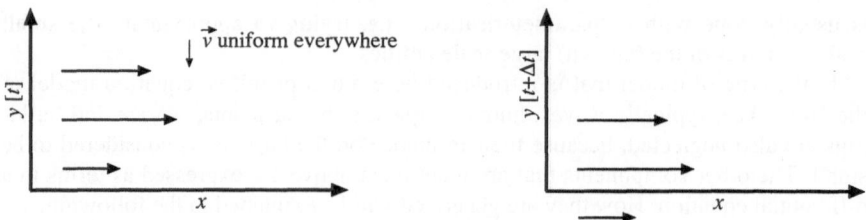

Fig. 5.4. A field of east-west (u-)velocities is advected south with a negative v-velocity

5.3.2   Wind and Sea Bottom

Wind and sea bottom, although seemingly quite different processes, affect the ocean currents basically in the same way, namely in the form of a frictional stress. This frictional stress τ originates from two surfaces in contact with each other that move at different speeds. The stress that is exerted depends on the vertical velocity difference on the one hand, and on the viscosity μ on the other hand:

$$\tau_x = \mu \quad \partial u / \partial z$$

$$\tau_y = \mu \quad \partial v / \partial z$$

The vertical velocity gradient comes in, because if both surfaces move at exactly the same speed, there is no friction, and as speeds become more different, it increases.

The viscosity is a measure that indicates how deep the influence will penetrate. If you imagine a wooden spatula that is drawn over a surface of (rather viscous) honey, it is quite obvious that much more honey will be moved than water, if the same spatula is drawn over a surface of (less viscous) water in the same way.

One way to parameterise turbulence is to express it in the same way as a viscosity. The rationale behind this way of thinking is that turbulence – as viscosity – increases the depth of wind influence in the same way as the advective terms in the equations of motion work: whatever stress is exerted at the surface is transported down into the water column by the small scale velocities. This of course happens in all three dimensions and thus the viscosity terms appear in all three equations.

5.3.3   The Coriolis Force

The Coriolis force results from the fact that the earth is a rotating sphere and deflects every horizontal movement of water on the northern hemisphere to the right (and on the southern hemisphere to the left). To imagine how this works, picture a particle sitting on the equator. It moves – as the earth does – at a speed of about 40 000 km/day (the length of the equator goes round once per day) around the axis of the earth. As it moves north, the underlying earth moves at a slower speed, because the zonal circumference of the earth becomes smaller (it is zero at the North Pole). The particle is thus faster than the underlying earth and "overtakes" it – moving east relative to the earth, i.e. to the right. Analogously, a particle starting south from the North Pole will be slower than the underlying earth and seem to move west – deflected to the right again. For a zonal movement this is not as easy to see, but the same is true.

For the equations, this effect means that a positive v velocity (pointing north) induces a positive u velocity (pointing east), and a positive v velocity induces a negative u velocity (and vice versa). Thus, two more terms enter the two horizontal equations:

u-eqn.: $+2\Omega \ sin(\phi) \ v$
v-eqn.: $-2\Omega \ sin(\phi) \ u$

where Ω is the angular velocity of the earth and ϕ is the latitude. In general, $2\Omega \ sin\phi$ is abbreviated as f and called the Coriolis parameter.

Two additional terms that result from the mathematical derivation of the Coriolis force are usually neglected:

The influence of w on horizontal motion, because the vertical velocities are much smaller than the horizontal ones; and the influence of u on the vertical motion, because in the vertical direction gravity and pressure dominate the regime (whereas horizontally, generally no large forces are at work).

5.3.4   River Runoff, Solar Heating, Surface Cooling

River runoff, solar heating and surface cooling all primarily influence the ocean currents by changing the density ρ of sea water that is largely determined by temperature and salinity. Horizontal density differences create an "internal" pressure gradient, e.g. in east-west direction $\partial I(\rho)/\partial x$. Figure 5.5 shows how this works: imagine a bowl with a wall in the middle, one half full of olive oil (with relatively low density), the other half full of red vinegar (with relatively high density). Because the red vinegar is heavier than the same volume of olive oil, there is a pressure gradient in the bowl, and when the wall is taken out, the liquids will rearrange so that the olive oil sits on top of the vinegar.

These internal pressure gradients exist in the same way in sea water. The density differences are just much smaller, and the Coriolis force prevents such a straightforward readjustment on the ocean basin scale.

5.3.5   Sea Surface Elevation

Another contribution to the pressure gradient comes from the sea surface elevation. In much the same way as in the case of different densities, a higher water column on one side of the wall in a bowl (see Fig. 5.6) leads to more weight on

Pressure gradient

Internal:

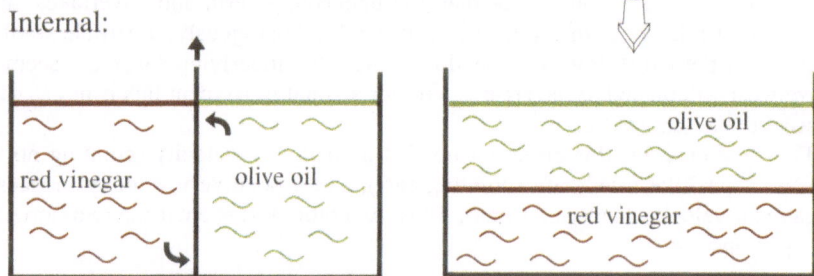

Fig. 5.5. Internal pressure gradient: the water column on the left hand side is heavier than that on the right. A pressure gradient is created that leads to readjustment when the wall is removed

Pressure gradient

External:

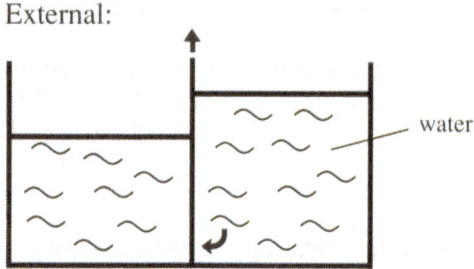

Fig. 5.6. External pressure gradient: again, the water column on the right hand side is heavier and the removal of the wall will lead to readjustment

one side and a readjustment as soon as the wall is taken out. A gradient in sea surface elevation (as for instance caused by the tides) is called the external pressure gradient, $\partial E(\zeta)/\partial x$ (cf. Backhaus 1985).

Both internal and external pressure gradients are added together, and have to be considered per unit mass for the equations, so that the terms

$$1/\rho \quad \partial\big(E(\zeta) + I(\rho)\big)/\partial x$$

$$1/\rho \quad \partial\big(E(\zeta) + I(\rho)\big)/\partial y$$

$$1/\rho \quad \partial\big(E(\zeta) + I(\rho)\big)/\partial z$$

have to be added to the equations. In the vertical, pressure gradient and gravitation are largest by far, so that all the other terms are generally neglected.

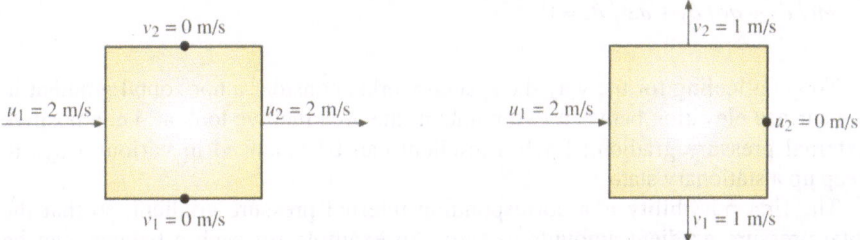

Fig. 5.7. In a two-dimensional box of given volume, inflow and outflow must balance: this can be true in just one direction (top), or the *u*- and *v*-velocities can balance each other

5.3.6 Continuity

One more equation is needed to fully determine the system of equations and there is one more thing we know about the dynamics of the sea: mass is preserved at any time. As a simplification, that is not quite correct but in a first order approximation, we assume sea water to be incompressible (this is true for fresh water, but not quite for saline water), so that continuity of mass and continuity of volume become equivalent. For a given volume we then know that inflow and outflow must balance, as sketched Fig. 5.7.

In terms of a differential equation, this can be expressed as:

$$\partial u \, / \, \partial x + \partial v \, / \, \partial y + \partial w \, / \, \partial z = 0.$$

The change of *u* in x-direction must be balanced by corresponding changes of *v* and *w* in y- and z-directions, respectively.

5.3.7 The Resulting Equations

Put together, all the terms we have created in the preceding sections result in a system of equations that is fully determined. They are called the *Navier-Stokes-Equations*, and most models of the ocean are based on them in some way or another (as can be seen in some of the other contributions to this volume). In the same way that we have considered the components we thought relevant, additional terms can be included that might matter under special circumstances. What we end up with in this system are the following equations:

$$du \, / \, dt + f \quad v + 1 / \rho \quad \partial \big(E(\varsigma) + I(\rho) \big) / \, \partial x = \tau_x$$

$$dv \, / \, dt - f \quad u + 1 / \rho \quad \partial \big(E(\varsigma) + I(\rho) \big) / \, \partial y = \tau_y$$

$$1 / \rho \quad \partial \big(E(\varsigma) + I(\rho) \big) / \, \partial z = g$$

$$\partial u \,/\, \partial x + \partial v \,/\, \partial y + \partial w \,/\, \partial z = 0$$

To get a feeling for the way the system works, consider a horizontal gradient in sea surface elevation between two points in the area that we look at, i.e. a positive external pressure gradient: Such a gradient can be balanced in various ways to keep up a stationary state.

The first possibility is a corresponding internal pressure gradient, so that the total pressure gradient amounts to zero. An example for such a balance can be found in the North Sea: The water along the Norwegian coast is less saline than that in the central and northern North Sea and just this difference in salinity leads to sea levels differences in the order of 10 cm, with higher sea levels at the coast, where the water is less dense (Damm 1997).

Another way to achieve a balance is a wind stress τ_x or τ_y that acts against a gradient in surface elevation. This comes to effect in a storm surge. In the German Bight that is shaped like a funnel open to the north-west, strong northwest winds that persist for a couple of days induce extreme sea levels and often considerable damage. In a strong storm surge, an elevation of 3 m above normal can occur due to wind stress (e.g. Flather 1999).

A third possible balancing term is the velocity term: a sea surface elevation at one place (if it is not balanced by any of the other terms) leads to a change in the velocities, i.e. water flows from the high elevation area to the surrounding places with lower sea surface elevation. In the North Sea, for instance, the tides work in this way, as the North Sea is too small to be attracted by the gravitation of the moon directly. So the tides are really only a wave that comes in from the North Atlantic. In this case, the Coriolis term that links the u and v equations becomes important: a strictly zonal change in the sea surface elevations will induce a change in u and via the Coriolis term produce an imbalance in the v equation. Then the whole system will start to oscillate, which is what usually happens in the real ocean (e.g. Hansen 1952).

5.3.8 Difficulties

There are a few difficulties inherent in this way of describing the ocean circulation. First, the density at every grid point has to be determined, as we have seen that it influences the velocities. However, usually not enough observations are available. Additionally, temperature and salinity, which determine the density, are transported by the currents. This problem is usually solved by adding equations for temperature and salinity, and the equation of state that gives the density from temperature and salinity (e.g. Pohlmann 1996).

The second (and grave) problem is the fact that the equations in this form cannot be solved analytically. There are basically two ways out: either one has to leave out most terms, further reducing the system to a balance of the two or three most important ones. Or the equations can be treated numerically. The first possibility gives exact solutions, but only for a very much simplified view of the ocean. With the second method, more processes can be taken into account, but the exactness of the solution has to be sacrificed: for a numerical treatment, the ocean has to be divided into a finite number of boxes, thus transferring a smoothly distributed field into a step function. Nevertheless, with improving computer performance, the numerical method has had a vast development since its first steps in the 1950s.

From this resolution problem, the difficulty of turbulence closure arises: in a finite grid, there are always scales that are not resolved, but have to be taken into account in some way. The parameterisation with the eddy viscosity concept that was mentioned above is just a rough approximation. More sophisticated methods have been constructed; one of the more widely applied ones is the k-ε model, where another two equations are introduced: one equation for the turbulent kinetic energy k, and one for its dispersion, ε (see e.g. Rodi 1980).

5.3.9 Boundary Conditions

The equations describe the **change** of the entities in time and space (x,y,z,t). To solve such a system of equations, for every entity there must be a neighbourhood from which the gradient can be determined. Because we are going to discuss the numerical rather than the analytical solution here, in the following it is assumed that the model area is divided into grid boxes. Then, there are two possibilities: using cyclic boundary conditions is one, where the "last" grid box is the neighbour for the first one. Otherwise, all along the boundaries, values have to be specified that provide the necessary gradients for the interior.

In the vertical, usually boundary values are specified, e.g. wind and bottom stresses and surface heating. In the horizontal, both are possible: when the global ocean is modeled, cyclic boundary conditions can be used naturally (with slight problems at the poles). In regional applications, boundary values have to be prescribed, typically to include the tides (in the form of surface elevations or velocities) and the density profile of the ambient ocean. Finally, in the time domain, initial conditions are given to provide a boundary condition at one end of the time scale. In one dimension, one end can be left open to let the system evolve, and that is usually the development in time. It would, however, in principle also be possible to prescribe all the boundaries (then including the final state at every grid point!) except for the eastern end of the domain and then see how the "initial" state in the west proceeds through the area.

5.3.10 Simplifications

With this step, not yet looking at the discretisation problem, we already lost a large part of nature, namely all the processes that are not expressed in the equations. Typical processes that are not part of the model are e.g. thermal expansion (in regional applications), salt anomalies that would enter through the boundaries, the influence of vegetation, or land sinking and rising. Some of these can be important for special applications and must then be taken care of. It is thus essential to know which processes are represented in a model.

5.4 From Differential Equations to a Numerical Representation

Now that we have the differential equations that describe the system we want to simulate, we have to transfer them into a form that makes it feasible to solve them. As mentioned before, for an analytical solution, drastic simplifications would be necessary, so we proceed here towards a numerical solution. To show the principle but keep things fairly simple, we consider the advection equation

$$\partial I/\partial t = -c\ \partial I/\partial x$$

that describes the advection of the concentration of a substance $I(x,t)$, e.g. blue ink, in a one-dimensional system (spatial dimension: x) with a velocity c. To avoid boundary conditions, we assume the model area to be a circular channel, with ink in a distribution as given Fig. 5.8.

Of course, the numerical representation of the full equations of motion will be more complex, but the basic principle is similar.

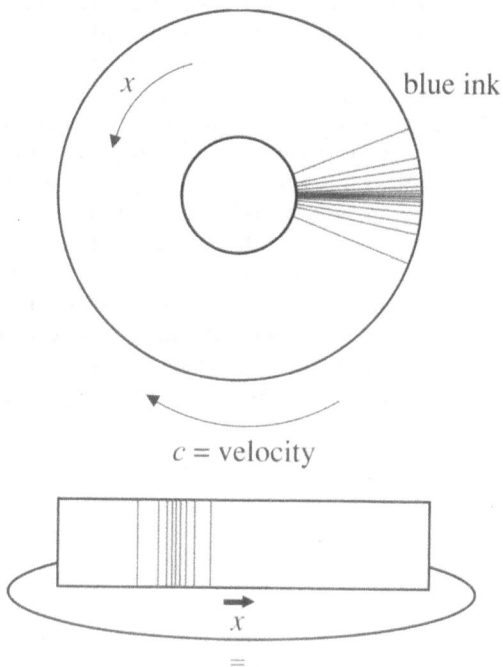

Fig. 5.8. Model setup: a circular channel without variation across-channel is assumed to be filled with water, moving with the velocity c. A blob of ink, that is slightly dispersed, is situated at one position in the channel (top). The circular channel can be thought of as a straight line, and the two ends are identified with each other (bottom)

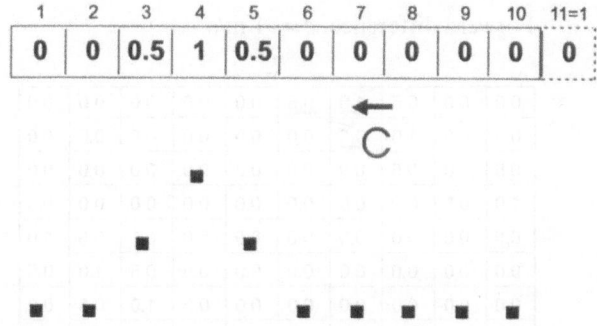

Fig. 5.9. Discretised form of the model setup: the channel is divided into ten boxes; each box is given a value for the concentration of ink. For the formulation of the boundary condition for $x = 10$, a new box $x = 11$ is introduced that has the ink concentration value of the box $x = 1$

5.4.1 Discretisation

The model domain has to be divided into a finite number of grid boxes (here: 10) and each box is assumed to be 1 m long in x-direction ($\Delta x = 1$ m). In the time domain, the time step is taken to be $\Delta t = 1$ s and the velocity $c = -1$ m/s, where the negative sign of the velocity means that it proceeds opposite to the direction in that x increases. The amount of ink in the boxes is taken to be 1 unit for $x = 4$, 0.5 for $x \in \{3.5\}$ and 0 for the other boxes (Fig. 5.9 for a sketch).

In the next step, the partial derivatives are replaced by difference terms in the following way, called "forward difference" in time and space:

$$\partial I / \partial t \rightarrow \frac{I(x, t + \Delta t) - I(x, t)}{\Delta t}$$

$$\partial I / \partial x \rightarrow \frac{I(x + \Delta x, t) - I(x, t)}{\Delta x}$$

Instead, a "backward difference" $\partial I / \partial x \rightarrow I(x, t) - I(x - \Delta x, t) / \Delta x$ or a "central difference" $\partial I / \partial x \rightarrow I(x + \Delta x, t) - I(x - \Delta x, t) / 2\Delta x$ can be used. The backward difference does not make sense for the time derivative, because what we are solving for is $I(x, t + \Delta t)$. Using the forward differences, the advection equation in discretised form then reads:

$$I(x, t + \Delta t) = -c \left(\frac{I(x + \Delta x, t) - I(x, t)}{\Delta x} \right) \Delta t + I(x, t)$$

In a forward difference, the gradient in the point x is estimated from the value at x and the value at $x + \Delta x$ (and in backward or central differences analogously). This might or might not be a good approximation.

Forward difference. $c = -1$ m/s

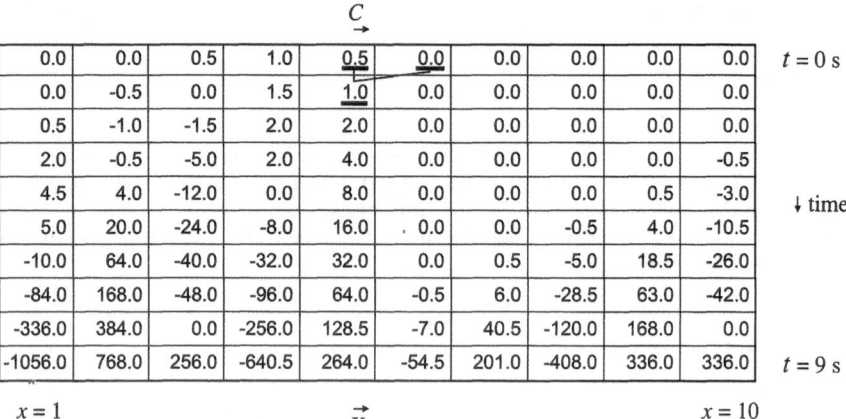

0.0	0.0	0.5	1.0	0.5	0.0	0.0	0.0	0.0	0.0	$t = 0$ s
0.0	0.5	1.0	0.5	0.0	0.0	0.0	0.0	0.0	0.0	
0.5	1.0	0.5	0.0	0.0	0.0	0.0	0.0	0.0	0.0	
1.0	0.5	0.0	0.0	0.0	0.0	0.0	0.0	0.0	0.5	
0.5	0.0	0.0	0.0	0.0	0.0	0.0	0.0	0.5	1.0	↓ time
0.0	0.0	0.0	0.0	0.0	0.0	0.0	0.5	1.0	0.5	
0.0	0.0	0.0	0.0	0.0	0.0	0.5	1.0	0.5	0.0	
0.0	0.0	0.0	0.0	0.0	0.5	1.0	0.5	0.0	0.0	
0.0	0.0	0.0	0.0	0.5	1.0	0.5	0.0	0.0	0.0	
0.0	0.0	0.0	0.5	1.0	0.5	0.0	0.0	0.0	0.0	$t = 9$ s

$x = 1$ $\overrightarrow{X}$ $x = 10$

Fig. 5.10a. Forward differences: the value of $x = 4$ at time $t = 2$ is calculated from the value at $x = 4$, $t = 1$ and the gradient between $x = 4$ and $x = 5$ at $t = 1$

5.4.2 Simulation

This equation is now used in a simulation. Fig. 5.10 shows the development in time for the case that was introduced above. This works exactly in the way we want it to work: the ink is transported once around the channel in 10 time-steps,

Forward difference. $c = +1$ m/s

0.0	0.0	0.5	1.0	0.5	0.0	0.0	0.0	0.0	0.0	$t = 0$ s
0.0	-0.5	0.0	1.5	1.0	0.0	0.0	0.0	0.0	0.0	
0.5	-1.0	-1.5	2.0	2.0	0.0	0.0	0.0	0.0	0.0	
2.0	-0.5	-5.0	2.0	4.0	0.0	0.0	0.0	0.0	-0.5	
4.5	4.0	-12.0	0.0	8.0	0.0	0.0	0.0	0.5	-3.0	↓ time
5.0	20.0	-24.0	-8.0	16.0	. 0.0	0.0	-0.5	4.0	-10.5	
-10.0	64.0	-40.0	-32.0	32.0	0.0	0.5	-5.0	18.5	-26.0	
-84.0	168.0	-48.0	-96.0	64.0	-0.5	6.0	-28.5	63.0	-42.0	
-336.0	384.0	0.0	-256.0	128.5	-7.0	40.5	-120.0	168.0	0.0	
-1056.0	768.0	256.0	-640.5	264.0	-54.5	201.0	-408.0	336.0	336.0	$t = 9$ s

$x = 1$ $\overrightarrow{X}$ $x = 10$

Fig. 5.10b. Forward differences (velocity positive in x-direction: the value of $x = 4$ at time $t = 2$ is still calculated from the value at $x = 4$, $t = 1$ and the gradient between $x = 4$ and $x = 5$ at $t = 1$, but the direction of the velocity is reversed

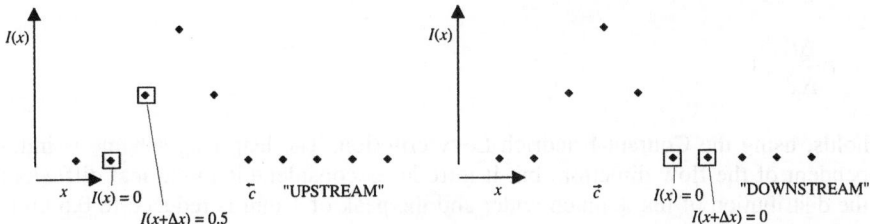

Fig. 5.11. When the velocity is negative, the gradient is taken from the box in question and the one upstream. This is appropriate, because the ink will be transported from the box upstream to the one in question with the velocity (left). When the velocity is positive, the gradient is taken from the box in question and the one downstream. Here, the gradient is 0, so the value $I(x)$ does not change. However, there is a peak of concentration approaching (from the other side) and it should have changed (right)

and the distribution does not change. However, this is due to the fact that the direction of the velocity is opposite to that of the difference taken. In such a case, the gradient is estimated from the value at x and that at the next point upstream. This makes sense and transports the ink correctly.

Using a forward difference in the case of a positive velocity ($c = +1$ m/s) results in numerical instability (cf. Fig. 5.10b). In this case the estimate must go wrong because it takes into account the point that does not matter for transport (see also Fig. 5.11.).

Finally, a method called the "leap-frog" scheme with a velocity of $c = 0.5$ m/s is presented in Fig. 5.12: it uses central differences in time and space. It is numeri-

$x = 1$			$c = 0.5$m/s					$x = 10$		
0.0	0.0	0.5	1.0	0.5	0.0	0.0	0.0	0.0	0.0	$t = 1$ s
0.0	-0.1	0.3	1.0	0.8	0.1	0.0	0.0	0.0	0.0	
0.1	-0.1	-0.1	0.8	0.9	0.4	0.1	0.0	0.0	0.0	
0.1	-0.1	-0.2	0.5	0.9	0.6	0.2	0.0	0.0	0.0	
0.1	0.0	-0.3	0.2	0.9	0.8	0.3	0.1	0.0	0.0	
0.2	0.2	-0.3	-0.1	0.7	0.9	0.5	0.2	0.1	-0.1	
0.1	0.2	-0.2	-0.3	0.4	0.8	0.7	0.3	0.2	-0.1	
0.1	0.3	0.0	-0.4	0.1	0.7	0.8	0.4	0.3	-0.1	
0.0	0.3	0.2	-0.3	-0.2	0.5	0.8	0.6	0.4	0.0	
0.0	0.2	0.3	-0.3	-0.3	0.2	0.7	0.6	0.6	0.1	$t = 10$ s

Numerically stable for $c \dfrac{\Delta t}{\Delta x} < 1$

Fig. 5.12. Leapfrog scheme: the value of $x = 4$ at time $t = 3$ is calculated from the value at $x = 4$, $t = 1$ and the gradient between $x = 3$ and $x = 5$ at $t = 2$

cally stable if

$$c \frac{\Delta t}{\Delta x} \leq 1$$

holds, using the Courant-Friedrich-Levy criterion. The leap-frog scheme is independent of the flow direction, but it introduces considerable numerical diffusion: the distribution of ink is much wider and the peak of 1 unit is reduced to 0.6 units within 10 timesteps.

One method that works better than the simple forward, backward or central differences is the "upstream" scheme, where the direction in that the differences are taken is determined from the velocities, so that the upstream case like Fig. 5.10a is always used.

Also, all of this was an "explicit" treatment of the equation. With "implicit" methods, where the value at the new time step appears on both sides of the equation, the strict Courant-Friedrich-Levy criterion need not quite be met. However, then a system of equations must be solved simultaneously, making the procedure more time consuming.

5.4.3 Simplifications

With discretisation, we lose another part of nature, additionally to what we neglected when setting up the system of equations: the small scales, in time as well as space. Thus, whatever is smaller or faster than the grid size or the time step, respectively, is not accounted for explicitly. For instance, topographic detail is always neglected, often even on islands like Heligoland in the German Bight; fronts between two water masses are smeared out or disappear altogether (e.g. Langenberg and Pohlmann 1994). In the vertical, stratification might be neglected, or gradients might appear much smaller than measured because of the resolution and finally, in the time domain, for global models the time step is often too large to calculate the tides explicitly.

Although this problem should decrease with decreasing grid sizes (which in practice come with faster and better computers), it is inherent in the numerical method and can not be solved.

5.5 Summary

There is a large variety of mathematical models of the environmental system, and some are treated with more detail in other contributions to this school.

The approach to modeling the ocean that was presented in detail here is a two-step-approach: first, those aspects of nature that are to be modeled are expressed in terms of a finite number of differential equations and boundary conditions. Here, we lose the openness of the system, i.e. the unlimited number of feedbacks and influences that characterise reality. In the second step, the equations are discretised and transformed into a numerical representation. This destroys the accuracy of the solution, i.e. the result that is calculated does not solve the equations exactly, but only approximately.

It is essential to keep those simplifications in mind in order to give a meaningful interpretation of the results.

Acknowledgments

I would like to thank G. Targonski and J. Sündermann, who taught me to understand and use mathematical models.

Chapter 6
Physical Modeling of Flow and Dispersion

by Michael Schatzmann

6.1 Introduction

Numerical models are commonly grouped into different categories depending on the scales they are applied to. If we do the same with physical models (wind tunnel and water tank experiments), we would assign them to the urban and street scale or, in other words, to the obstacle resolving scale.

This scale is of particular importance for industrial and urban air pollution problems, with the exception of tall stacks, the emissions occur here mainly within or shortly above the canopy layer, i.e., the zone where the atmospheric flow is heavily disturbed by buildings and other obstructions and where people live. Due to the combined effects of natural wind variability and obstacles, local concentrations differ substantially over short distances. Models applicable to such problems must be able to resolve multiple obstacles as well as the numerous eddies shed by them.

Although there has been significant progress in the development of microscale meteorological models over the last few years, they are still limited in scope, mainly due to a lack of computational power. The largest machines presently available accommodate only about 100 x 100 x 50 grid cells for problems as we discuss them here, which restricts the applicability of these codes to rather idealised geometric configurations. Moreover, such models require new schemes for the parameterisation of sub-grid scale turbulence, which remains to be developed. Therefore, the simulation of in-canopy layer dispersion processes is still the domain of physical models.

The methods applied in physical modeling will subsequently be demonstrated with an example of practical relevance, the dispersion of accidentally released flammable or toxic gases. The consequences of such potential accidents which might occur in chemical industry need to be assessed, and if these consequences appear to be inacceptable, actions to improve the safety of the plant must be taken. Other examples can be found in, e.g., Schatzmann et al. (1998).

6.2 Properties of Wind-Tunnel Boundary Layers

The reliability of results from physical model studies depends strongly on the quality of the mean and turbulence characteristics of the simulated boundary layer.

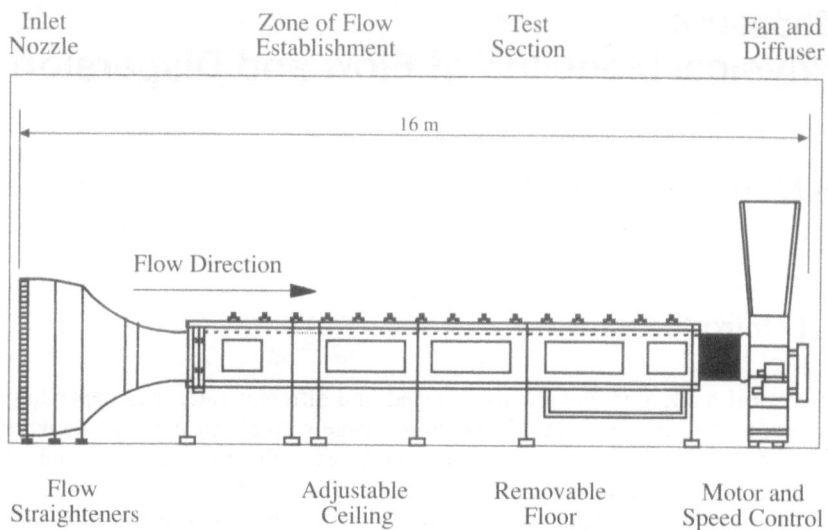

Fig. 6.1. Neutrally stratified boundary-layer wind-tunnel of Hamburg University. Size of cross section: 1.5 m x 1.0 m

Subsequently, some results of such characteristics measured in a wind tunnel boundary layer will be presented and compared with those of the atmospheric boundary layer.

Most wind tunnels which are utilised to generate the small scale boundary layer have a design similar to that sketched in Fig. 6.1. They consist of an inlet nozzle with flow straighteners, a long flow establishment section, a test section, and a fan driven by a speed controlled electric motor. Most tunnels work with suction in order to keep any disturbance of the flow inside the tunnel to a minimum. An adjustable ceiling is used to establish a zero-pressure-gradient boundary layer. The change of wind direction with height or density stratification can not be simulated in such a tunnel. However, there are other tunnel designs available, which are specialised just on the simulation of stable and unstable boundary layers or even elevated inversions (see, e.g., Schatzmann et al. 1995).

As in dispersion calculations for licensing purposes, the vertical velocity profile of the wind tunnel boundary layer is usually approximated by the power law

$$\frac{u(z)}{u(\delta)} = \left(\frac{z}{\delta}\right)^n \tag{6.1}$$

with $u(\delta)$ the velocity at boundary layer height δ.

Desired power law exponents n can be achieved by use of specific combinations of vortex generators (Irwin 1981) and artifical roughness elements distributed over the bottom of the flow establishment section (Fig. 6.2). Figure 6.3 gives an example of a mean vertical velocity profile in logarithmic presentation obtained in the tunnel. In this particular case, a boundary layer thickness of $\delta = 0.4$ m and a profile exponent of $n = 0.16$ was obtained.

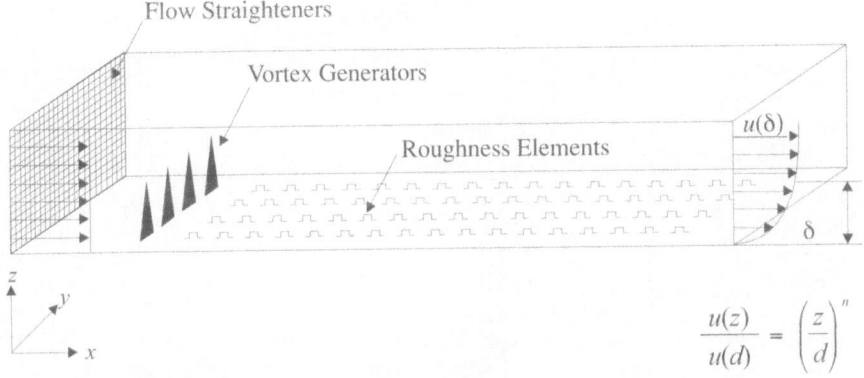

Fig. 6.2. Generation of a neutrally-stratified boundary-layer wind profile by use of vortex generators and artifical roughness elements

To obtain model/prototype similarity, not only the mean but also the turbulence properties of the boundary layer have to be matched. Since the Reynolds number

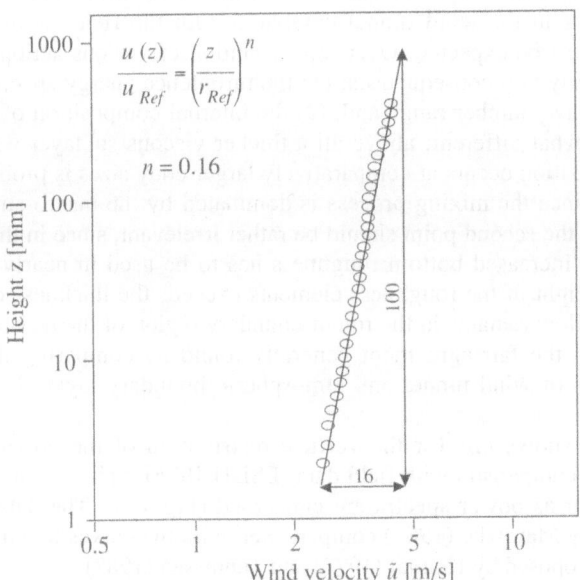

Fig. 6.3. Vertical profile of wind velocity at the end of the zone of flow establishment section

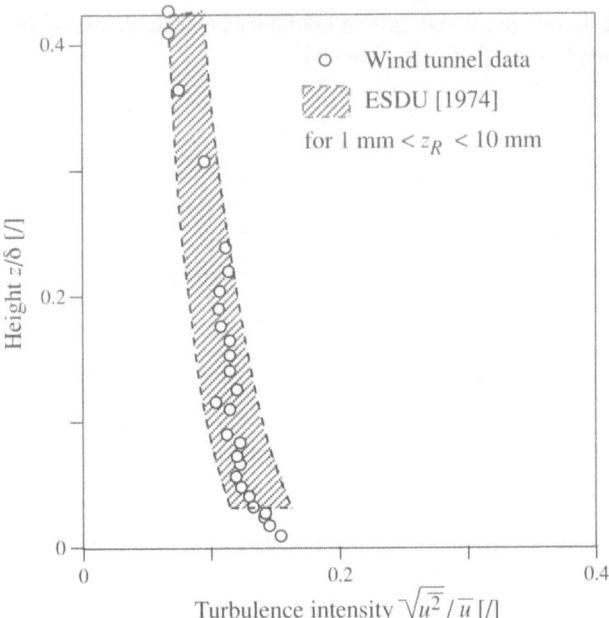

Fig. 6.4. Vertical turbulence intensity profile in comparison with ESDU (1974) field data

representing the ratio of inertial and viscous forces takes on a value some orders of magnitude less in the wind tunnel compared with the field, some fundamental differences must be expected. Over representation of viscous action in the model results in mainly two consequences: (1) the turbulence energy spectrum is cut off in the large wave number range and, (2) the internal composition of the boundary layer is somewhat different; above all a thicker viscous sublayer will occur. The fact that dissipation occurs at comparatively larger eddy sizes is probably of minor importance, since the mixing process is dominated by the macro-structure of turbulence. Also the second point should be rather irrelevant, since in the wind tunnel an artificially increased bottom roughness has to be used in nearly any case. As long as the height of the roughness elements exceeds the thickness of the viscous sublayer, the flow remains in the rough-boundary region of the resistance diagram. Consequently, the fair agreement generally found by comparing the turbulence characteristics of wind tunnel and atmospheric boundary layers is not very surprising.

Figure 6.4 shows this for the vertical distribution of the horizontal velocity fluctuation in comparison with field data (ESDU 1974). Fair agreement is likewise obtained as far as power spectra are concerned (Fig. 6.5). The data taken in our wind tunnel by Marotzke (1987) compare well with the curves based on field data, which were proposed by Kaimal (1972) and Teunissen (1982).

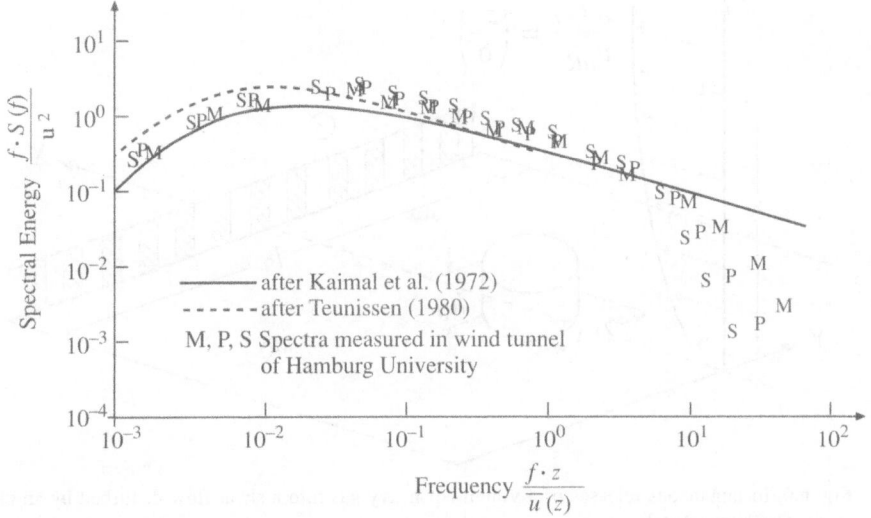

Fig. 6.5. Power spectra measured in the wind tunnel in comparison with spectra based on field data according to Kaimal et al. (1972) and Teunissen (1982) (from Marotzke 1991)

6.3 Dimensional Analysis

Certain similarity requirements must be fulfilled in order to transfer results from small-scale wind tunnel experiments to prototype scale. These similarity laws are usually obtained by dimensional analysis, a method that makes use of the fact that physical equations must be dimensionally homogeneous and hence the parameters occurring therein can only appear in certain combinations.

The derivation of similarity laws is subsequently outlined in the example of an instantaneous heavy gas release as it was realised in the Thorney Island field tests (Mc Quaid 1984). Figure 6.6 shows the situation. A volume V_o of heavy gas with density ρ_o and viscosity μ_o is released into an atmospheric boundary-layer flow characterised by a power law wind profile with exponent n and a boundary layer thickness δ. The initial shape of the cloud is unknown in reality, so we assume for simplicity a cylinder with height h_o and radius ρ_o.

The properties of the surrounding air are described by ρ_a and μ_a, z_a characterises the surface roughness, and $\bar{u}_{aR}$ the velocity in a given reference height. The spread of the cloud is affected by several obstacles. Their geometry is fully defined by the length scales $l_1, ..., l_n$ and the porosity Θ.

We restrict ourselves to the so-called gravity spreading zone which is dominated by the (negative) buoyancy of the cloud and extends down to a few percent of the initial gas concentration. Within this zone, density differences between the cloud and its environment are much larger than those in the ambient flow. Therefore, the influence of ambient stratification on the spread of the cloud is small and can be neglected. With this restriction in mind, the local dilution $\chi = \Delta c / \Delta c_o$ (with

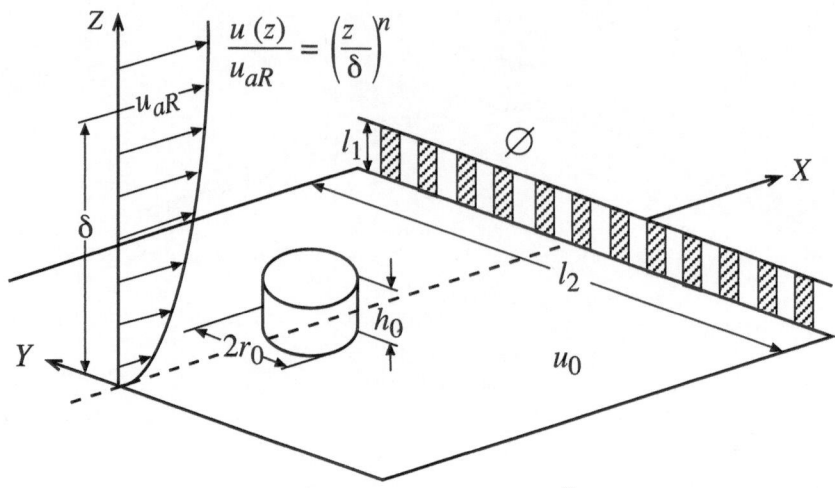

Fig. 6.6. Instantaneous releases of a volume of heavy gas into a shear flow disturbed by an obstacle. Definition sketch

Δc and Δc_o the local and initial concentration excess above ambient, respectively) depends on the following variables

$$\chi = f_1\left(t, x, y, z, g, \rho_o, \mu_o, V_o, r_o, z_R, \rho_a, \mu_a, \ \bar{u}_{aR}, \delta, l_1, \ldots, l_n, \Theta\right) \tag{6.2}$$

Applying dimensional considerations, a characteristic length scale, time scale and velocity scale

$$L_{ci} = V_0^{1/3} \tag{6.3a}$$

$$T_{ci} = \left(\frac{L_{ci}}{g'}\right)^{1/2} \tag{6.3b}$$

$$U_{ci} = \left(L_{ci} g'\right)^{1/2} \tag{6.3c}$$

can be defined with $g' = g \Delta \rho_o / \rho_a$ the 'effective' gravity acceleration and $\Delta \rho_o = \rho_o - \rho_a$.

The subscripts c and i indicate that these scales are (c)haracteristic variables for an (i)nstantaneous gas release. Nondimensionalisation of Eq. 6.2 results in

$$\chi = f_2\left(\frac{t}{T_{ci}}, \frac{x}{L_{ci}}, \frac{y}{L_{ci}}, \frac{z}{L_{ci}}, \frac{\Delta \rho_o}{\rho_a}, \frac{L_{ci} U_{ci}}{\mu_o / \rho_o}, \frac{L_{ci} \bar{u}_{aR}}{\mu_a / \rho_a}, \frac{r_o}{L_{ci}}\right. \tag{6.4}$$

$$\left.\left(\frac{z_R}{L_{ci}}, \frac{\bar{u}_{aR}}{U_{ci}}, \frac{\delta}{L_{ci}}, \frac{l_1}{L_{ci}}, \ldots, \frac{l_n}{L_{ci}}, \Theta\right)\right.$$

Equation 6.4 indicates that the local dilution χ as a function of the normalised time t/T_{ci} in both the model and full-scale takes on identical values at locations determined by the coordinates x/L_{ci}, y/L_{ci} and z/L_{ci}, provided that all the remaining dimensionless parameters on the right hand side of this equation can also be matched in the wind tunnel experiment. These parameters are: the density excess ratio of the initial cloud, a cloud Reynolds number, and an ambient Reynolds number, the aspect ratio of the initial cloud r_o/L_{ci}, the roughness parameter z_R/L_{ci}, the velocity ratio $\bar{u}_{aR}/U_{ci}$ which has the quality of a Froude number (as can be seen when U_{ci} is replaced by Eq. 6.3c), the nondimensionalised mixing height δ/L_{ci}, the dimensionless length scales, l_1/L_{ci} to l_n/L_{ci}, and the porosity Θ.

It should be noted that both the dimensional analysis and the wind-tunnel experiments do not include those processes leading to the formation of the initial cloud (burst of a tank, evaporation of a liquified gas). A possible ignition of the cloud in case the gas is flammable is also excluded from the analysis. We assume that the source is fully described by an appropriately chosen initial density and initial volume and concentrate solely on cloud dispersion.

In the case of a continuous release (Fig. 6.7), the initial volume V_o is replaced by the initial volume flux $\dot{V}_o$. Since V_o contains the basic units length and time, dimensional analysis results in the characteristic scales

$$L_{cc} = \left(\frac{V_0^2}{g'}\right)^{1/5} \quad T_{cc} = \left(\frac{\dot{V}_o}{g'^3}\right)^{1/5} \quad U_{cc} = \left(\dot{V}_o g'^2\right)^{1/5} \tag{6.5}$$

and into the non-dimensionalised variables

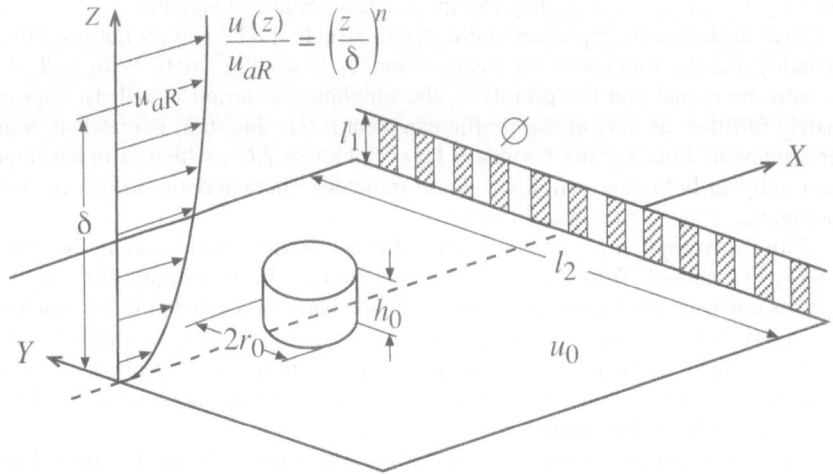

Fig. 6.7. Continuous heavy gas releases from a ground level source into a shear flow disturbed by an obstacle. Definition sketch

$$\chi = f_3\left(\frac{t}{T_{cc}}, \frac{x}{L_{cc}}, \frac{y}{L_{cc}}, \frac{z}{L_{cc}}, \frac{\Delta\rho_o}{\rho_a}, \frac{L_{cc}U_{cc}}{\mu_o/\rho_o}, \frac{L_{cc}\bar{u}_{aR}}{\mu_a/\rho_a}, \right.$$
$$\left. \frac{r_o}{L_{cc}}, \frac{z_R}{L_{cc}}, \frac{\bar{u}_{aR}}{U_{cc}}, \frac{\delta}{L_{cc}}, \frac{l_1}{L_{cc}}, ..., \frac{l_n}{L_{cc}}, \Theta\right) \tag{6.6}$$

The subscript cc identifies a (c)haracteristic variable in the case of a (c)ontinuous release .

Note that our definition of similarity parameters differs from that proposed by Britter et al. (1988). The main difference lies in the fact that Britter employed the ambient wind speed to form the characteristic scales. Since the zero ambient wind case is excluded from the analysis thereby, we did not follow this approach.

6.4 Matching of Similarity Requirements

In a small-scale model it is neither possible nor necessary to match all the similarity numbers listed in Eqs. 6.4 and 6.6 to prototype values. The t/T_c -terms disappear when we restrict our analysis to ensemble mean values of excess concentration maxima (instantaneous release) or to suitably defined time mean values (steady continuous release). In general, the Reynolds numbers (terms 6 and 7 on the right hand side of the equations) are significantly smaller in a physical model than in the field. This has, however, only marginal implications as long as the Reynolds numbers are kept above a certain critical value. For the cloud Reynolds number, $L_{ci}U_{ci} / (\mu_0/\rho_0)$, this value is about 400, according to Janssen (1981). For continuous spills, the critical value of the Reynolds number depends on the specific release conditions. For low-momentum ground-releases which are considered here, $L_{cc}U_{cc} /(\mu_0/\rho_0)$ is, according to experience, a peripheral variable.

Since obstacles are represented directly in a scale model and do not need to be included into the roughness parameterisation, z_R is usually small. With $z_R/L_c \ll 1$ in both the model and the prototype, the roughness criterion is at least approximately fulfilled, as long as the profile exponent n (see Fig. 6.2) is matched. Similar arguments hold for the boundary layer thickness ΔL_{ci}, which is much larger than unity in both cases (Index c alone indicates characteristic values for both instantaneous and continuous releases).

With the exception of extreme values that we do not consider here, the aspect ratio r_0 / L_c should affect the cloud development only in the vicinity of the source. Further downstream details of the initial cloud shape are expected to become progressively less relevant. For the determination of lower flammability distances (LFD), defined as the distance from the source within which ignitable gas concentrations occur, it seems to be sufficient to characterise the cloud by its initial volume (or volume flux, respectively) alone.

With the provisions concerning the Reynolds number in mind, approximate similarity is to be expected by using an undistorted wind tunnel model and matching the density excess ratio and the velocity ratio. To prove this hypothesis, model/full-scale comparisons have been carried out (see Sect. 6.8).

6.5 Experiments

The dispersion experiments in the wind tunnel were carried out with sulphur hexafluoride (SF_6)/air mixtures. Instantaneous spills have been performed utilising a cylindrical container with a volume of 450 cm^3. At the time of release, the side wall of the container was abruptly retracted into the wind tunnel floor. Continuous releases were realised through an orifice mounted flash into the tunnel floor. Volume fluxes of 100 l h^{-1} to 500 l h^{-1} have been emitted. The vertical momentum of the discharge was small.

In a first series of experiments, instantaneous and continuous releases in unobstructed, flat terrain were carried out. The importance of the individual similarity parameters were investigated, and the results were compared with those known from field experiments. Subsequently, the experiments were repeated with 25 different obstacle arrays.

Concentration fluctuation measurements were carried out by using specially designed aspirated hot wire probes (König et al. 1987). To keep disturbances of the flow to a minimum, only the tops of the probe inlet tubes reached into the test section. The bodies of the probes were mounted below the wind tunnel floor.

The time resolution of the probes was about 20 Hz. SF_6-concentrations down to about 0.1% were detectable. Concentration versus time traces were measured simultaneously at eight ground-level locations. A transient recorder with a corresponding number of channels was available for digitising and processing the probe signals.

6.6 Variation of Similarity Parameters

Accidents involving heavy gases have already occurred many times in the past. Summarising accounts of such accidents as they were published by, e.g., the German Environmental Protection Agency, Umweltbundesamt (UBA 1983) or the Association for Technical Monitoring, TÜV Norddeutschland e.V. (TÜV 1982), indicate a large diversity among the cases reported on. They differ with respect to the following aspects:

- type of gas release (instantaneous or continuous),
- amount of the released gas,
- physical properties of the gas,
- atmospheric conditions,
- characteristics of the canopy around the source.

To cover the heavy gas dispersion problem in its full complexity leads to a huge experimental programme, which is hardly tractable anymore. To reduce the number of experiments to a minimum, some of the similarity parameters in Eqs. 6.3 and 6.5 were varied in a systematic way in order to quantify their importance for the dispersion process.

The similarity requirement to keep both the excess density ratio $\Delta\rho_o/\Delta\rho_a$ and the densimetric Froude number (identical with the velocity ratio $\bar{u}_{aR}/U_c$) the same in model and prototype, has the consequence of inconveniently and sometimes even impractically low wind velocities in the tunnel. To match both parameters in a physical model corresponds to the application of a mathematical model utilising the full, non-Boussinesq-approximated equations. On the other hand, relaxing the excess density ratio requirement and using the Froude number as the sole scaling parameter for buoyancy effects would correspond to a Boussinesq-approximated mathematical model. Since the introduction of the Boussinesq-approximation is a common tool in mathematical modeling, it was appealing to check its consequences also in a physical model by distorting the density ratio.

Controversy still surrounds the use of distorted density scaling. Whereas concentration versus time traces measured by van Heugten and Duijm (1984) show lower peak concentrations but extended time durations of cloud passage, pertinent experiments carried out by Hall et al. (1982) and Havens (1984) indicate only differences within normal experimental scatter. On the other hand, in simulating the China Lake field trials in a density distorted model, Neff and Meroney (1982) found the heavy gas clouds moving generally too slowly in the wind tunnel.

Our own results show exactly the same trend as reported by Neff and Meroney (see Fig. 6.8 for instantaneous releases). When we repeat a certain experiment by varying the density excess ratio but keeping the densimetric Froude number the same, we notice clearly a cloud speed reduction in proportion to the density distortion. Local peak concentrations, however, seem to be only marginally effected. Since a shift in cloud arrival and departure times does not change the lower flam-

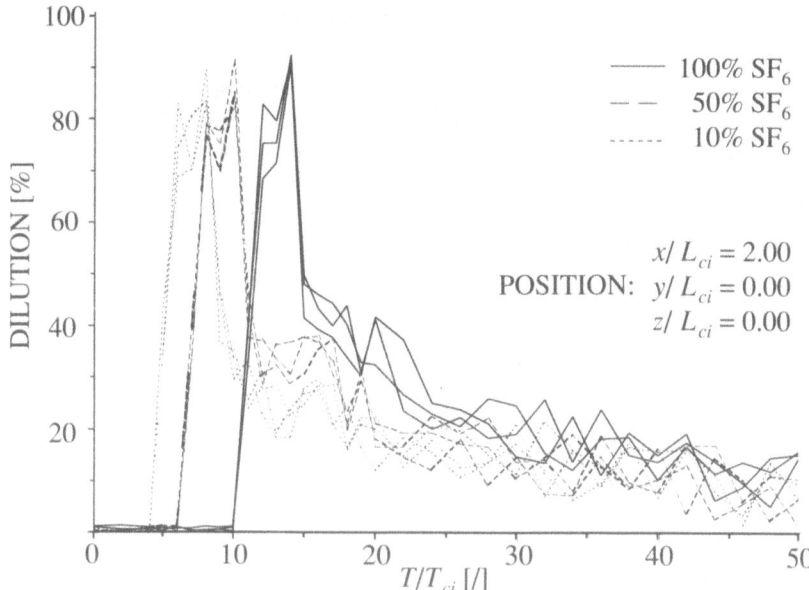

Fig. 6.8. Concentration versus time traces for instantaneously released heavy gas clouds at a fixed position for different initial densities (flat terrain, $\bar{u}_{aR}(L_{ci} = U_{ci})$)

mability distances, and thus the area within which the gas/air mixture is hazardous, the subsequent experiments with instantaneous spills can be carried out with the Froude number as the essential scaling parameter, the density ratio remaining as a free parameter.

Results from experiments with continuous spills exhibit a similar feature. Changing the density excess ratio $\Delta\rho_o/\rho_a$ by a factor up to ten has only marginal implications on ground level concentration decay (Fig. 6.9). Especially in the

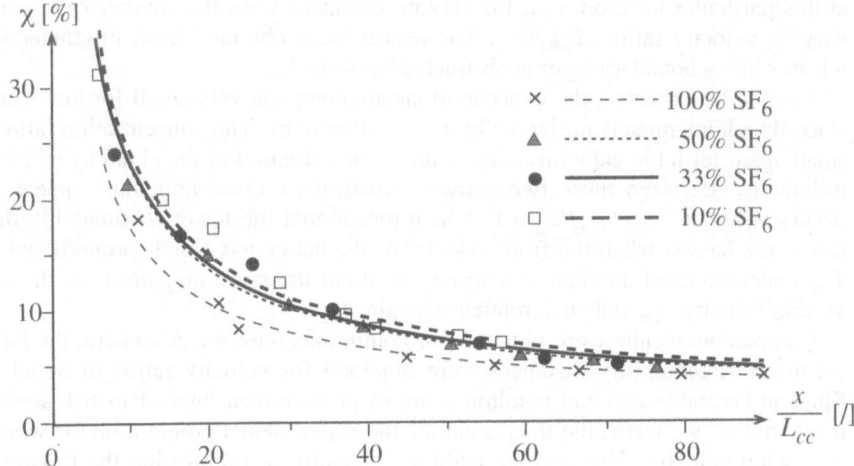

Fig. 6.9. Concentration decay with distance as a function of density excess ratios for continuous releases (flat terrain, $\bar{u}_{aR}\ (L_{cc}) = U_{cc}, z\ /L = L_{cc} = y\ /L_{cc} = 0$

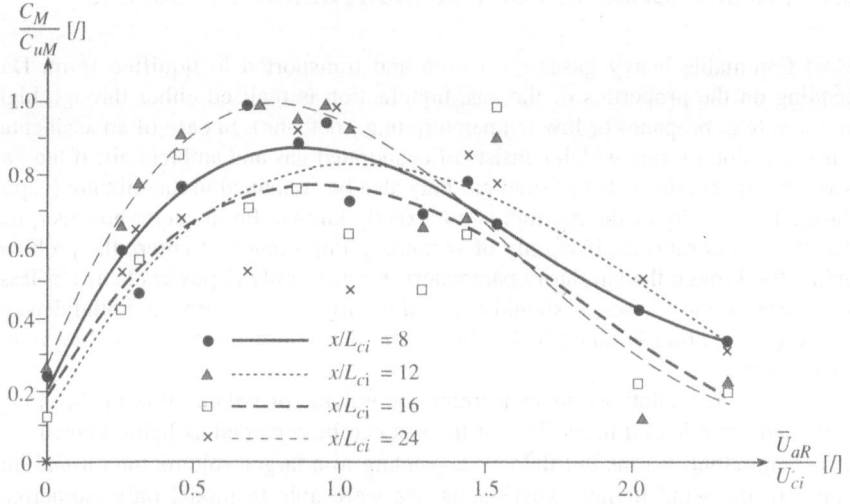

Fig. 6.10. Normalised concentration maxima c_M/c_{uM} as a function of wind velocity ratio $\bar{u}_{aR}/U_{ci}$ at several ground-level positions for instantaneously released heavy gas clouds

range of lower flammability concentrations, the differences are well within normal experimental scatter. In conclusion, Froude number modeling is appropriate also in the case of continuous releases. The requirement of identical density excess ratios can be relaxed.

Risk prevention strategies are usually based on worst-case assessments. To discover unfavourable conditions, the wind velocity ratio $\bar{u}_{aR}/U_{ci}$ was systematically varied and the concentration field at ground level monitored. Figure 6.10 shows the functional dependence of local maximum ground level concentrations (ensemble means), c_M, normalised with the maximum of all concentration maxima at this particular location, c_{uM}, for varying distances from the source, x/L_{ci}, and varying velocity ratios, $\bar{u}_{aR}/U_{ci}$. The results were obtained from instantaneous releases into a boundary layer unobstructed by obstacles.

As is to be expected, the concentrations are comparatively small for low wind, since the cloud spreads horizontally into all directions. The concentration ratio is small again for large velocities, due to the intense dilution of the cloud by ambient turbulence. Between these two extreme situations a maximum must appear. It occurs at about $u_{aR}(L_{ci})/U_{ci} = 1$, which means that the lower flammability distances are largest when the front velocity of the heavy gas cloud (proportional to U_{ci} under stagnant ambient conditions) of about the same magnitude as the advection velocity $\bar{u}_{aR}$ (taken at reference height L_{ci}).

Comparable results were obtained for continuous releases. Also here, the largest lower flammability distances were obtained for velocity ratios of about 1. Since unfavorable ambient conditions are of predominant interest in risk assessment studies, we were able to concentrate the experimental program on the worst-case wind velocity. This had the additional benefit of diminishing the Reynolds number problem, since $\bar{u}_{aR}(L_c) = U_c$ is usually a high wind velocity.

6.7 Parameterisation of Thermodynamic Processes

Most flammable heavy gases are stored and transported in liquified form. Depending on the properties of the gas, liquefaction is realised either through high pressure (e.g. propane) or low temperature (e.g. methane). In case of an accidental release, a cloud forms which consists of evaporated gas and ambient air. If the gas was pressurised, droplets of liquid gas may also be contained in the mixture (vapor flash). The density of the mixture is not exactly known. Fortunately, however, the density excess ratio itself is only of secondary importance. It enters the problem indirectly through the similarity parameters, but here only at powers to 1/2 or less. Therefore, in most cases, it should not lead to large errors when the initial density of the initial cloud is taken to be simply the density of the gas at boiling point temperature.

Another uncertainty exists concerning the volume or volume flux of the cloud. Although the released mass flow of the gas can be regarded as being known, the dispersing cloud warms up, thereby expanding to a larger volume than it had initially. In the wind tunnel experiments, we were able to model only isothermal clouds. For compensation, we increased the initial volume V_o (or volume flux V_o) to the value it takes on when warmed up to ambient temperature.

To determine the sensitivity of lower flammability distances (LFD) on these source parameters, some tests with varying mass fluxes of pressurised propane released continuously in unobstructed, flat terrain were carried out. In the first case, the density was set to $\rho_0 = 1.87$ kg / m³, thereby assuming that the heavy gas warms up to ambient temperature immediately after the spill, with a corresponding increase in V_o. In the second scenario the density of propane at boiling point temperature $\rho_0 = 2.36$ kg / m³ was used and the increase in volume neglected.

Figure 6.11 shows for the two cases the lower flammability distance (based on mean values of the lower flammability limit for propane, i.e. 2.1 % by volume) as a function of the spill rate for unfavorable wind conditions $(u_{aR}(L_{cc}) = U_{cc})$. Although extreme situations were assumed, the LFDs differ by only about 20 %, which is not much in the context of a safety analysis.

At first glance, it is surprising that the lower flammability distance decreases when the gas density increases. The reason for this is that the lateral spread is more intense for the heavier cloud with the consequence of a reduced LFD-value. Since, on the other hand, the neutrally-dense cloud does not give the largest flammability distance, there must be a worst-case density ratio. According to our experience, it is in the range between $0 \le \Delta\rho_o/\rho_a \le 0.5$; i.e. most heavy gases of practical interest have density ratios above that range.

Another unknown quantity in the formulation of the source term is the amount of air mixed into the cloud during cloud formation. For risk analysis applications it seems to be appropriate to neglect mixing processes at the source. The release of a (then smaller) volume of pure gas, in comparison with the mixture containing an identical mass of heavy gas, should always lead to the larger flammability distance, and therefore to a conservative estimate. We were also able to prove this assumption experimentally.

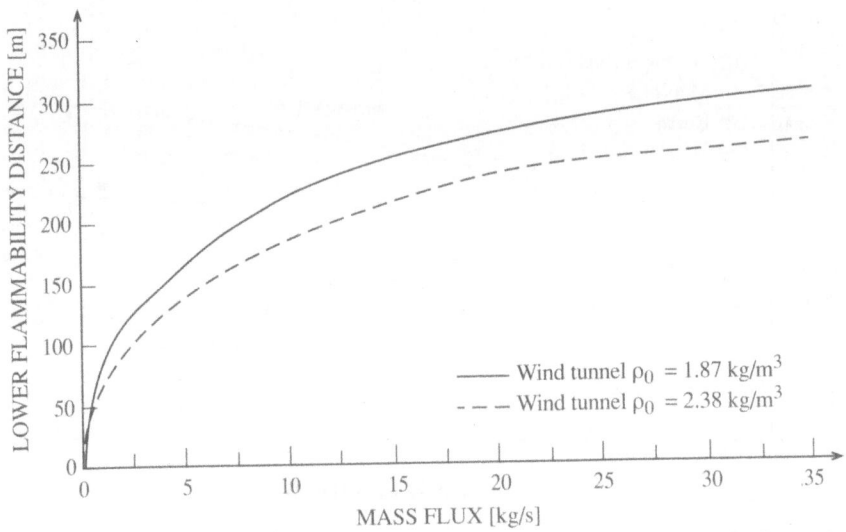

Fig. 6.11. Sensitivity of lower flammability distances for fixed mass fluxes but different release scenarios (see text)

6.8 Small-Scale/Full-Scale Comparisons

Wind tunnel validation experiments have been carried out using field data from continuous release tests with methane (Burro and Maplin Sands experiments) and propane (TÜV Hamburg and Maplin Sands experiments). The field data cover different source conditions from nearly momentum-free spills of deep-cooled liquified propane and methane over flash releases to jet-like discharges of pressurised propane. The spill rates varied over a wide range. The experiments were carried out both over land and over water at varying ambient wind and stratification conditions. For details see Kóopmann et al. (1982), Puttock et al. (1982, 1985) and Heinrich et al. (1986).

In Fig. 6.12 and 6.13 the lower flammability distances in the field trials are presented as a function of spill rate (symbols). They are compared with those from the wind tunnel tests (curves) which have been carried out with pure SF_6 and at an ambient wind speed of $\bar{u}_{aR}(L_{cc}) = U_{cc}$. As in the field, the test area was flat and unobstructed. For comparison, the initial density in the large scale experiments was assumed to be the gas density at boiling point temperature; the initial volume flux V_o was increased to the value it takes on at ambient temperature; and finally, entrainment of ambient air into the cloud during cloud formation was neglected.

Figure 6.12 shows the comparison for LNG. Due to concentration fluctuations within the cloud, lower flammability distances based on ground-level mean concentration values (dashed line) and maximum concentration values (continuous line) were determined. According to our theory, the maximum curve (defined as mean + 2RMS) should envelop all field data, which were similarly based on concen centrations maxima but not necessarily taken under unfavorable ambient wind conditions.

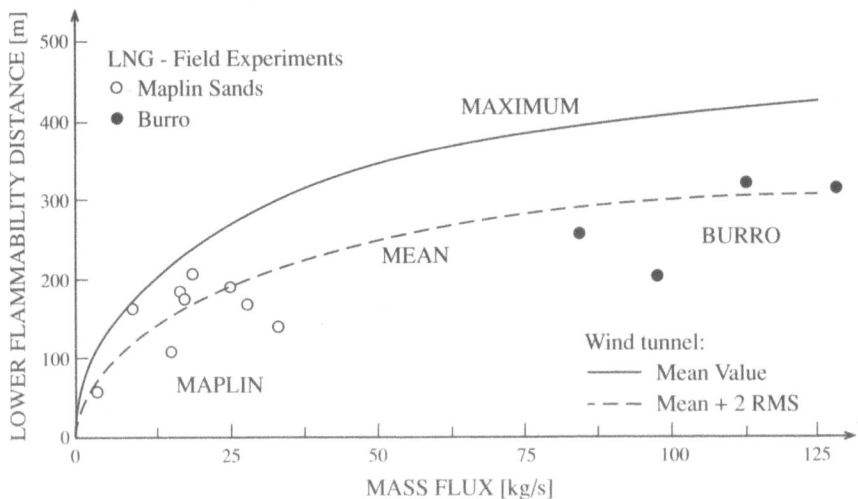

Fig. 6.12. Comparison of data obtained in the Maplin Sands and Burro field trials with results from the wind tunnel experiments

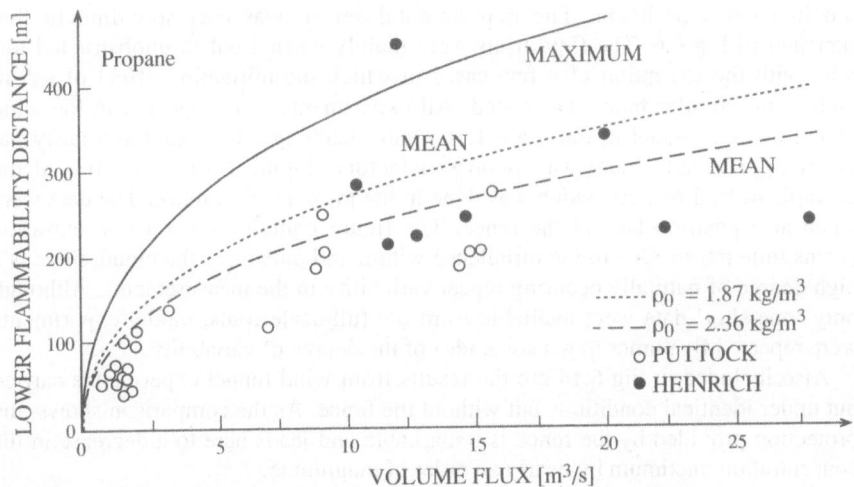

Fig. 6.13. Comparison of data obtained in the Maplin Sands and TÜV Hamburg field trials with results from wind tunnel experiments

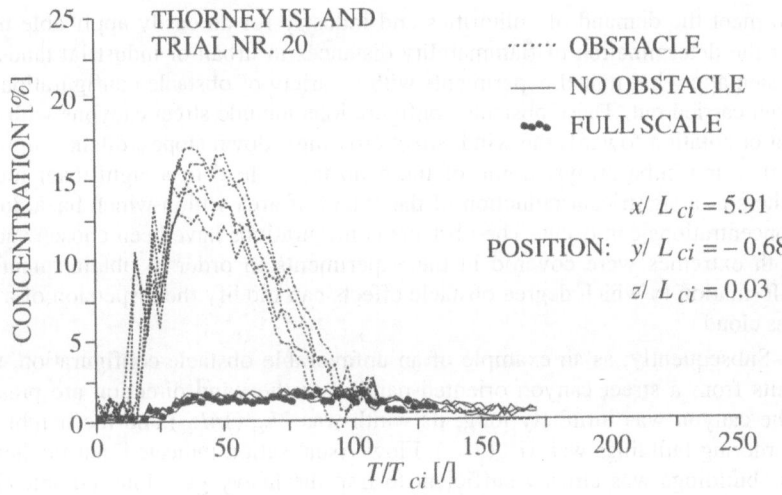

Fig. 6.14. Small-scale/full-scale comparison of Thorney Island experiment No. 20. The potential concentration versus time traces in the absence of the fence are also included in the figure

Fair agreement is likewise obtained in the comparison for propane spills (Fig. 6.13). Small-scale/full-scale comparisons with instantaneously released heavy gases have been carried out using field data obtained in the HSE heavy gas dispersion field trials at Thorney Island, U.K., during 1982/83 (McQuaid 1984). Using a fold-away tent, 2000 m³ of denser-than-air gas was instantaneously released in each experiment. The experimental set-up was corresponding to that sketched in Fig 6.6. The field trials were mainly carried out in unobstructed terrain, with the exception of a few cases in which the mitigating effect of a 5 m high, semi-circular fence was tested. All experiments were repeated in the wind tunnel using a model in the scale 1:165 and matching all relevant similarity parameters. The agreement was again satisfactory. Figure 6.14 shows this at the example of trial No. 20, which was done in the presence of a fence. The data were taken at a position behind the fence. The figure contains several concentration versus time traces. Due to the turbulence within and outside of the cloud, there is a high degree of naturally occuring repeat variability in the measurements. Although only 'one-shot'-data were available from the full-scale trials, model experiments were repeated five times to get some idea of the degree of variability.

Also included in Fig 6.14 are the results from wind tunnel experiments carried out under identical conditions but without the fence. As the comparison shows, the protection provided by the fence is remarkable and leads here to a decrease in the concentration maximum by nearly an order of magnitude.

6.9 Investigation of Obstacle Effects

To meet the demand of authorities and industry for an easily applicable method for the determination of flammability distances in urban or industrial landscapes, systematic wind tunnel experiments with a variety of obstacle configurations have been carried out. These obstacle configurations include street canyons with different orientation towards the wind, street crossings, down slopes, ditches, mitigating walls, and cube arrays. Some of these obstacles lead to a significant increase, others to a significant reduction of the extent of area within which hazardous gas concentrations can occur. The obstacle configurations have been chosen such that both extremes were covered in the experiments in order to obtain quantitative information to which degree obstacle effects can modify the dispersion of a heavy gas cloud.

Subsequently, as an example of an unfavorable obstacle configuration, the results from a street canyon oriented parallel to the wind direction are presented. The canyon was infinitely long; its width was $2L_{ci}$ ($14L_{cc}$) and the height of the bordering buildings was $1L_{ci}$ ($7L_{cc}$). Flow visualisation indicated that the height of the buildings was already sufficient to trap the heavy gas cloud completely. Instantaneous as well as continuous releases were realised. The wind velocity $\bar{u}_{aR}$ measured at height L_c was chosen to be the characteristic wind speed U_c. Ground-level concentration versus time traces have been recorded at eight locations downstream from the source. Mean and maximum (mean +2 RMS) values of the local concentrations $\Delta c/\Delta c_0$ were determined by time averaging the concentration versus time traces (steady continuous releases) or ensemble averaging the excess concentration maxima obtained in 10 repeats of the same experiment (instantaneous releases), respectively.

The results obtained have been transformed into graphs, as presented in Fig. 6.15. They allow the direct determination of worst-case lower flammability distances as a function of the lower flammability concentration for all major flammable gases.

Assuming e.g., the instantaneous release of 2000 kg pressurised propane, the density of the gas at boiling point temperature (-42.1 ^{0}C) is $\rho_0 = 2.38$ kg m^{-3} and supposing the propane evaporates completely, the initial volume of the cloud is $V_0 = 1050$ m^3 (taking into account the expansion of the cloud due to temperature increase to ambient temperature during dispersion; see Sect. 6.7). According to Eq. 6.2, the characteristic length of the problem is $L_{ci} = 10.16$ m. Assuming the release of pure propane and considering that the lower flammability concentration (LFC) is 2.1 %, Figure 6.15 provides the lower flammability distances based on mean and maximum ground-level concentrations, $LFD_{mean} = 37\ L_{ci} = 376$ m and $LFD_{max} = 41\ L_{ci} = 417$ m, respectively.

If the same spill were to occur in flat, unobstructed terrain, according to our measurements, the corresponding LFD values would be 234 m and 305 m. In the

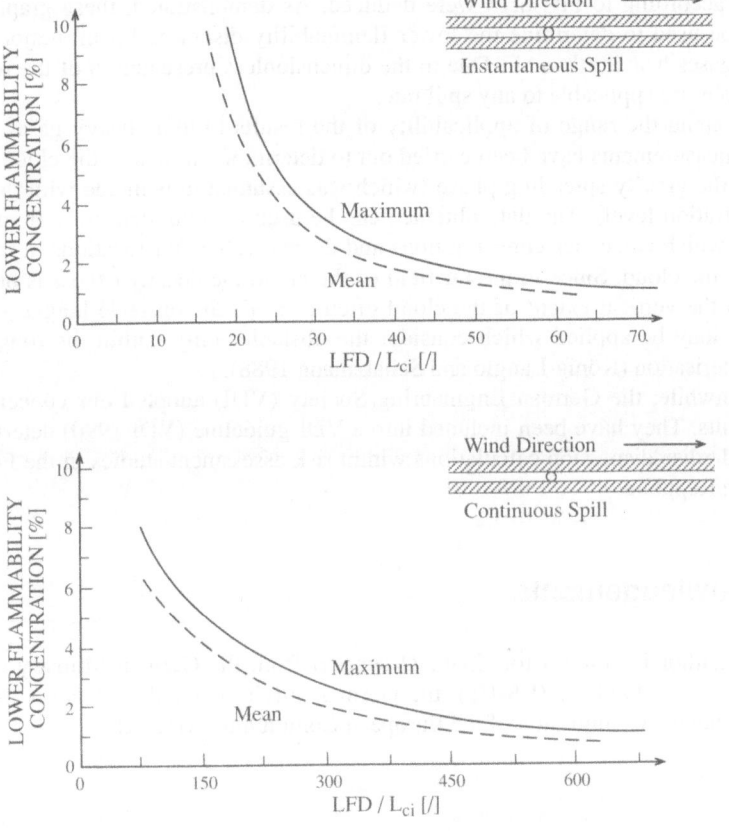

Fig. 6.15. Lower flammability distances for instantaneously (top) and continuously (bottom) released heavy gas for varying lower flammability concentrations LFC

case of a semicircular wall of 0.4 L_{ci} height located 4 L_{ci} downwind from the source, the pair of lower flammability values would decrease to 42 and 66 m, respectively.

The street canyon described above represents one of the most unfavourable obstacle configurations, whereas the semi-circular wall, on the other hand, proved to be one of the most effective mitigating structures in our experiments. The range within which lower flammability distances change due to obstacle effects is therefore known. Obstacle arrays of practical interest are likely to be between these two extremes. A first estimate of LFD-values can therefore also be obtained for those obstacle configurations which have not yet been covered in our investigation.

6.10 Conclusions

So far, 25 obstacle configurations have been realised in our wind tunnel experiments (see König 1987 and Marotzke and Schatzmann 1993). From the results, graphs according to Fig. 6.15 were deduced. As demonstrated, these graphs can easily be used to determine the lower flammability distances for all major flammable gases heavier than air. Due to the dimensionless presentation of the results, the graphs are applicable to any spill rate.

To extend the range of applicability of the results to toxic heavy gases, additional measurements have been carried out to determine the size of the cloud at the end of the gravity spreading phase (which was assumed to coincide with the 1 % concentration level). The data obtained can be used as input data for a numerical model, which calculates concentrations and dosage values for locations in the far-field of the cloud. Since in the far field of the cloud the density excess is negligible and the vertical extent of the cloud often exceeds the obstacle height, simple models may be applied which consider the obstacles only within the roughness parameterisation (König-Langlo and Schatzmann 1988).

Meanwhile, the German Engineering Society (VDI) adopted our concept and the results. They have been included into a VDI-guideline (VDI 1990) determined to standardise dispersion calculations within risk assessment studies in the Federal German Republic.

Acknowledgements

The author is grateful for financial support from the German Ministry of Research and Technology (BMFT), the German Environmental Protection Agency (UBA) and the Commission of the European Communities (DG XII).

Chapter 7
Conceptual Models for Ecology-Related Decisions

by Karl-Heinz van Bernem

Abstract

The high variability of complex biological interactions causes large statistical and systematical errors, which make it impossible to realistically reflect nature using complex simulation models. An adequate correspondence with reality, however, is a prerequisite when considering the tolerance of ecosystems to man-made disturbances and countermeasures.

Simplified conceptual models can be a very helpful tool when considering ecological topics in decision making (management) processes. Three examples are introduced which are assigned to the decision-fields "Environmental Sensitivity Indices (ESI)", "Environmental Risk Assessment (ERA)" and "Environmental Impact Assessment (EIA)". The selected examples are based on realistic scenarios concerning a particular environment at the German North Sea Coast, the "Wadden Sea". They imply models to define the sensitivity of corresponding coastal areas to oil-pollution, a conception to decide about the application of chemical dispersants as part of oil spill response measures, as well as a programme to monitor the effects of a pipeline landfall to sub- and intertidal communities.

7.1 Introduction

Ecology related decisions are not only in demand when estimating long term environmental effects, but considerably more frequent when estimating damages or changes following short term "man made" interferences.

Decisions of this kind presuppose the prediction of systemic reactions. They usually aim at a mitigation of intensity, duration and spatial extension of expected effects. As result ecology-related criteria are often achieved by including criteria concerning natural resource protection. The priorities chosen in each individual case are decided by a subjective evaluation in the frame of ecosystem-management (Starfield and Bleloch 1986).

Conceptions of models inevitably remain inductive in theoretical ecology because universal principles and laws are lacking to a large extent. Thus simulations of complex interactions in nature are bordered definitely by narrow limits. The term "ecosystem," often used but not exactly defined, contains various "subsystems" which show determined functional and structural connections. Certain connections, for example "predator/prey relations", "intra- and interspecific concur-rence," and "optimal reproduction" (to mention just a few) may be described more or less "well" (realistic for distinct systems) by empirical or dynamical mod-els. It is difficult and usually impossible, however, to simulate complex systems. In order to decrease systematic mistakes, the number of compartments taken into consideration has to be increased, but this, in turn, increases the statistical error. (Fig. 7.1) The latter is because of the high variability caused by biological events, often so large that clear statements are impeded.

A subjective selection of models in connection with the evaluation of decision criteria mentioned above and the state of the data (usually insufficient) aggravates or prohibits the application of complex "simulation-models" in short-term deci-sion-making processes.

On the other hand, fundamental processes and functions of soundly investigated "ecosystems" are quite well known and independent of biological fluctuations. Similar conditions are considered for key factors of certain functions and key spe-cies, which share the progress of processes. If we confine ourselves to such pa-rameters, and in further limitation, to a selection of those important to the ques-tioning, very simple models result. Simplistic conceptual models of this kind al-low the consistence of hypotheses and assumptions to be tested. They should al-ways be aimed at a concrete question, in which their structure is defined by the objective. Quality and availability of data play a minor role. As far as quantifica-tions are necessary, they can mostly be replaced by estimated or assigned values stemming from expert-knowledge (Starfield and Bleloch 1983).

The empirical possibility of validifying the "state of reality" of such models makes them particularly suitable for short term and general decision-processes.

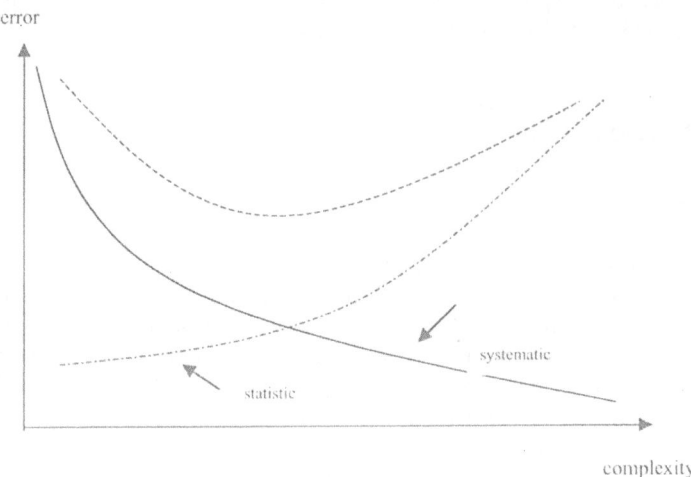

Fig. 7.1. Dependence of the systematic and statistic error on the complexity of the model. Total error expected = hatched line (modified following O'Neill, 1973 in Wissel 1989)

In the following, three examples of marine coastal management problems are introduced, which are assigned to the decision-fields "Environmental Sensitivity Indices (ESI)," "Ecological Risk Assessment (ERA)" and "Ecological Impact Assessment (EIA)."

"ESI" represent fundamental principles for the classification of coastal areas with respect to their different vulnerability to oil pollution; an "ERA" serves the management of unplanned incidents (ship accidents etc.) while an "EIA" is reserved for planned interventions such as harbour-constructions, river-dredging and the laying of pipelines. All cited examples refrain from using mathematical formulations, but for all of them the following prerequisites are essential: the conceptions have to be as simple as possible but nevertheless should be based on the best data and knowledge available, and they need validation by reality (otherwise a decision, made on the basis of these concepts, might be the last made by the operator on duty).

All selected examples stem from realistic scenarios concerning a particular environment at the North Sea Coast, the "Wadden Sea." Because each kind of ecological modeling is closely tied to the underlying systems involved, some characteristic properties of this coastal zone are sketched in brief.

7.2 The Wadden Sea – a Sensitive Environment

A contiguous region of tidal flats, barrier islands, alluvial terrestrial zones, and salt marshes, about 450 km long and up to 20 km wide, extends along the North Sea coast of Germany, the Netherlands and Denmark. This "Wadden Sea" is of enormous value as a cleansing site for the coastal water, a nursery for young fish, and a feeding and nesting ground for nearly all palaearctic species of wading birds and waterfowl. Predation is one of the most important processes. It keeps densities of the large burrowing infauna (organisms living below the sediment surface) below carrying capacity, thus positively influencing the amelioration of the sediment. The interactions between the constituent species on the tidal flats are not chaotic but they undoubtedly lack perfection. As a consequence, it is inherently impossible to arrive at clear-cut conclusions about their pattern. What can be said is simply that certain sequences of events are more likely to occur than others, but it is never possible to reject some other sequence once and for all.

The proximity of important shipping routes and ports is a permanent threat, especially to the German part of the region, which became a national park in 1985/86. Large quantities of petroleum, for example, which can be spread over wide areas by tides and winds, present not only the danger of temporary damage but also permanent harm, since oil, bound to the sediment, is released very slowly and can therefore repeatedly contaminate those parts of the tidal flats that have become free of the oil.

Aside from these accidents, planned measures like dredging, disposal and large scale construction mean a lasting stress to this environment.

7.3 The Environmental Sensitivity Index (ESI) for Wadden Sea Areas

Dicks and Wright (1989) summarised some fundamental aspects concerning "coastal sensitivity mapping." They emphasised the importance of environmental sensitivity as a valuable and useful concept in the control and management of industrial and urban development and in contingency planning. Foreknowledge of both sensitivity and vulnerability of habitats is essential to environmental planning and protection and in counteracting and minimising the impact of unplanned activities.

Plans for vulnerability or sensitivity indices were first developed among others by Gundlach and Hayes (1978). The following are underlying questions:

- Is it possible to define areas of different vulnerability to oil?
- Which factors determine this vulnerability?

They established a concept that was mainly based on spatial geomorphological parameters and designed for regions that are fundamentally different from the Wadden Sea Coast. In Fig. 7.2 the main parts of this concept are depicted. The degree of sensitivity of coastal areas is marked by distinct interactions, which take place if they are polluted by oil slicks: a high degree of wave energy will soon reduce the oil coverage of rocky shores and sandy beaches; the correspondingly high content of oxygen will guarantee a fast degradation of the oil. Low hydrological energy in sheltered coastal areas will cause longer-lasting coverage and penetration of sediments, in some cases leading to stable oil-layers in different sediment depths. Both coverage and penetration will hinder the oxygen exchange and so enlarge the persistence of pollution. The only biological factor in these assumptions, aside from microbial degradation processes, is the mortality of sensitive organisms. This factor is marked as damage without further differentiation (Fig. 7.2).

As a result of these conceptual assumptions, the questions mentioned above can be answered leading to the following kinds of habitats listed in increasing order of vulnerability.

1. Exposed rocky headlands
2. Erosive wave-cut platforms
3. Fine-grained sandy beaches
4. Coarse-grained sand beaches
5. Exposed tidal flats
6. Mixed sand and gravel beaches
7. Gravel beaches
8. Sheltered rocky coasts
9. Sheltered tidal flats
10. Salt marshes and mangrove swamp

Although this concept could be validated by comparing the results to real effects of several oil accidents, its applicability is limited. Areas which are geomor-

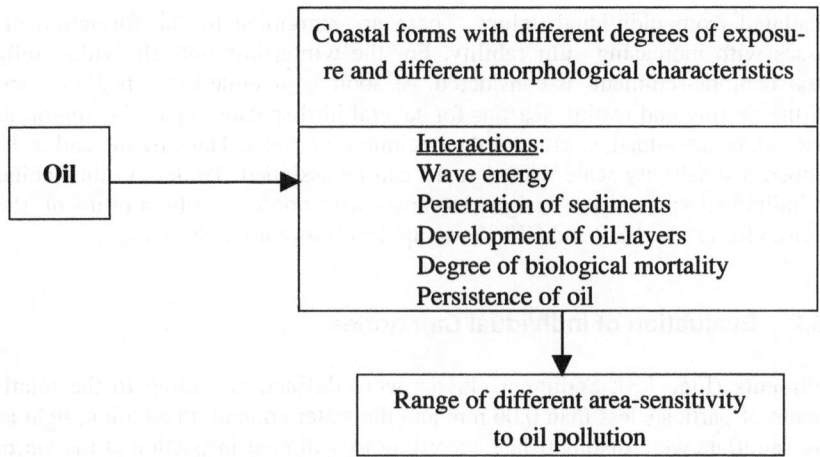

Fig. 7.2. Concept of spatial oil-sensitivity of coastal forms

phologically more or less homogenous (for example the Wadden Sea) contain only a few of the habitats depicted (No. 5, 9, 10), and a temporal differentiation is absent.

This situation made it essential to greatly modify the concept for the German North Sea coast, because a total protection of the wide spread habitats classified by this model was estimated as being impossible. The areas and seasons of this region for which special protection is required should be identified by including a greater number of ecological parameters. In order to do this, the results of many years of field and laboratory experiments could be used (a.o. Giere 1979; van Bernem 1982; Dörjes 1984; Farke et al. 1985). They show that the short and long-term consequences of oil pollution in Wadden Sea areas clearly depend on a much wider range of habitats affected, and are determined by both abiotic and biotic parameters as interrelationships among toxicity, turbation, and persistence.

Thus we can determine the following enlarged questions:

Is it possible to further differentiate areas in space and time which are morphologically similar by integrating ecological criteria?

Which fundamental ecological factors (and factors of natural resource protection) determine vulnerability to oil?

The development of an evaluation system of species and habitats with respect to their sensitivity to oil pollution should make it possible to obtain basic features to determine their vulnerability to oil in individual cases on the one hand, and to summarise them, with the necessary degree of simplification, for use in contingency planning, on the other. It should also make information available to the national park authorities and serve scientific purposes, i.e., in ecosystems research.

7.3.1 The Evaluation

All groups in the evaluation; sediment, benthos (organisms living on or in the sea bottom), fishes, shrimp populations, birds, and salt marshes were assigned indices, calculated from individual values. These are smoothed by the formation of 4 classes with increasing vulnerability. For the wintertime only the vulnerability class "benthos-sediment" is considered. As soon as juvenile fishes begin to arrive and the nesting and resting seasons for several bird species begin, the appropriate class values are added, starting at the beginning of April. Thus, by the end of November, a sensitivity scale from 1 to 12 can be assigned. To delimit the habitats, the individual values were assigned to areas with borders set by a protocol, standardised for use by EDP and GIS (Geographical Information Systems).

7.3.2   Evaluation of Individual Categories

Sediments (Fig. 7.3). Sediment classes were defined according to the relative amount of particles less than 0,06 mm and the water content. In addition, light and dark sandflats were distinguished according to sediment inspection at the various stations and evaluation of aerial images. Thus, class 1 referred to light sand flats, class 2 to dark sand flats, class 3 to tidal flats of sand and silt, and class 4 to mud flats.

Benthos (Fig. 7.3). When evaluating the vulnerability of the species and communities to petroleum, giving consideration to their significance in the ecosystem, the following categories were established: (1) physiological sensitivity, (2) ecological sensitivity, (3) importance as food, (4) metabolic importance, (5) capability of dispersal, and (6) duration of reproductive period. Within these categories, every species was assigned a weighted value ranging from 1 to 3, where 1 signified weak or minor, and 3 strong or high. An example is provided in the following evaluation of greater benthos organisms (macrofauna):

Physiological sensitivity was judged according to the experience gained by research in the field and laboratory. This scale includes the following levels:

1. Species with little change in abundance after exposure to petroleum.
2. Significant decline in abundance.
3. Very significant decline in abundance.

Ecological sensitivity can be determined by observing settlement patterns and food consumption:

1. Endobenthic sand dwellers, substrate feeders, or predators.
2. Sand flat dwellers that feed on the surface, nonfiltering inhabitants of mixed sand and mud flats, and residents of mud or sandy mud flats that can tolerate oxygen deficits.
3. Predators with tentacles, filter feeders, and species that live on the surface of mud flats.

According to Beukema (1981), the average weight of the biomass in the Dutch Wadden Sea amounts to 26,6 g ash-free dry weight per m^2. Almost 99 percent of this, 26,2 g, was accounted for by only 14 species. Relying on the average weights of the dominant species in the biomass, the *importance as food* of the macrofauna species was estimated according to the following scale:

1. Species with less than 0,1 g ash-free dry weight/m^2 that are rarely preyed upon.
2. Species with less than 0,1 g ash-free dry weight/m^2 that are frequently preyed upon by fish or birds.
3. Species with greater than 0,1 g ash-free dry weight/m^2.

Several species increase the oxygen supply in the sediment through their movements, while certain sediment and epistrate feeders have a controlling or destructive effect on populations. The following levels were used for the criterion *metabolisation of organic substances:*

1. Inhabitants of detritus-poor sand sediments or suspension feeding sessile species.

Categories	Species (examples)		
	macrofauna	meiofauna	microalgae
	Arenicola marina	*Harpacticus flexus*	*Achnanthes spec.*
physiological sensitivity	2	3	1
ecological sensitivity	1	3	1
importance as food	3	3	1
metabolisation capacity	3	1	2
dispersal capacity	2	1	3
duration of reproductive period	3	3	1
$\sum$	14	14	9
$\sum/n$	2.3	2.3	1.5

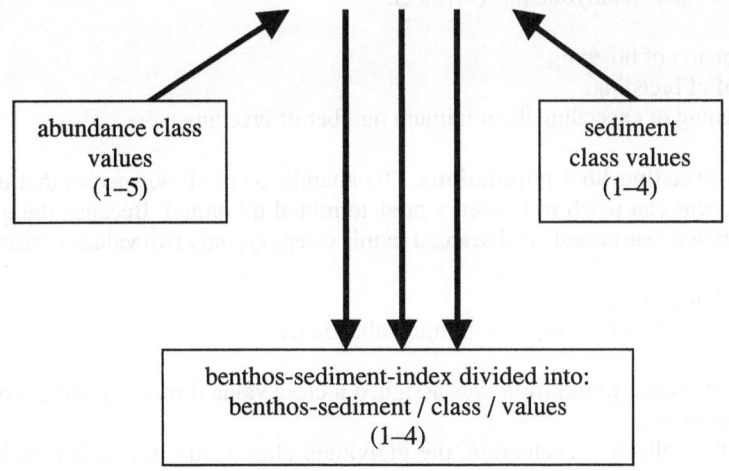

Fig. 7.3. Relationships among the values influencing the "benthos-sediment" complex

2. Species active in bioturbation with feeding habits that scarcely contribute to the breakdown of organic substances.
3. Substrate and epistrate feeders greatly active in bioturbation.

In the recolonisation of tidal flats, very mobile species and those with planktonic larval stages have an advantage. The following scale is used for evaluating the *dispersal capability* of species:

1. Actively swimming species with a life history that includes a planktonic larval stage.
2. Actively swimming species that develop in the benthos.
3. Species with a limited locomotory capability without planktonic larval stages.

Duration of the reproductive period is an additional criterion to consider. Species that produce larval stages for long periods of time have an advantage in being able to rapidly resettle unpopulated areas, as reflected in the following scale:

1. Reproduction throughout almost the entire year.
2. Reproduction during four to six months of the year.
3. Reproduction during one to three months of the year.

The arithmetic value for individual species and locations is calculated by regarding the estimated abundance of species which form communities at corresponding locations, according to van Bernem et al. (1989).

Breeding bird populations. Of the approximately 27 bird species present, 13 proved especially vulnerable. A "minimum breeding pair" number is assigned only to these species. The distinction between "suspicion of breeding" and "proof of breeding" provided other qualitative differences. All three qualities were assigned weighted values according to a model similar to those used for the groups "benthos" and "ichthyofauna" (=fishes):

1. Suspicion of breeding.
2. Proof of breeding.
3. Reaching or exceeding the minimum number of breeding pairs.

Non-breeding bird populations. The abundance of all 36 species that usually might occur can reach or exceed a predetermined minimum. Because the qualitative criterion "suspected" is discarded in this category, only two values remain:

1. Proof of presence.
2. Reaching or exceeding the minimum abundance.

The **moulting population** was assigned a class value directly, without arithmetic calculations.

For the collective evaluation, the individual class values for each month were added, and the sums were assigned new class values ranging from 1 to 4.

Saltmarshes. Based on research on the sensitivity of halophytes to petroleum (see Hershner et al. 1977; Baker 1983), evaluation criteria for the individual species were developed also comparably to those for the category "benthos".

Area of contact: Weighted values were 1 = small, 2 = medium, and 3 = large.

Position of the regeneration organs: 1 = underground (geophytes), 2 = more than 50 cm above the sediment, 3 = 10 to 50 cm above the sediment, 4 = 1 to 10 cm above the sediment, 5 = plants forming rosettes, and 6 = small (therophytes). Position of regenerative organs and location of new shoots after oil contamination were considered.

Physiological reaction: 1 = little, 2 = medium, 3 = strong, and 4 = very strong.

Regeneration: 1 = very rapid, 2 = rapid, 3 = medium, 4 = slow, and 5 = no regeneration.

Degree of endangerment: 1 = low, 2 = medium, and 3 = high. This is an evaluation of the exposure to harm in case of an accident. The main zone of endangerment is assumed to extend as far as the mid-tide level or a little beyond.

The points in the individual evaluation criteria are added and the sum is divided by the number of criteria (5). The quotient is assigned to one of eight classes corresponding to particular indicator values. For areas subject to grazing by cattle, this species sensitivity index is increased by two points, because contamination is intensified in the absence of protective leaf cover. Thus, a scale from 1 to 10 is produced. To give proper emphasis to the population density distribution of each species, a community vulnerabilty index was employed which is calculated according to the food value method of Klapp (1971):

$$CVi = \frac{\text{distribution of species} \times \text{vulnerability value}}{\text{total settlement density}}$$

The value for oil sensitivity of the specific kind of plant association (CVi) calculated by this method is then placed in one of four classes scaled according to vulnerability.

7.3.3   General Evaluation

By summarising the class values (1–4 each) achieved according to the examples demonstrated above, a maximum of 12 points can be obtained (benthos-sediment + fishes + birds + saltmarshes). In order to create a clear depiction, the results are demonstrated on different thematic maps (in Fig. 7.4 restricted to the complex "benthos-sediment + birds").

This vulnerability study was specially prepared for the Wadden Sea with regard to its ecological characteristics and established research data. It has been put into practice as part of the German Oil Spill Contingency Plan. As an operational model it demonstrates the possibility of establishing gradual differences of morphologically similar areas of oil pollution if we possess a sound knowledge of the oil-sensitivity of species/communities/habitats and their distribution in space and time. However we need monitoring at systems-dependent scales to adapt the evaluation to environmental changes.

Concerning the demands of other countries, we may conclude according to Dicks and Wright (1989): "We cannot advocate a uniform approach to map preparation. Like spills, the individual needs of a coastal area and its environment are unique, and the response and sensitivity maps benefit from tailoring to local demands. Nevertheless, our advice is to keep the maps simple, make sure they are designed for the user, and make it clear where expert guidance is needed."

The environmental sensitivity belongs to the level of precaution measures. On its basis, area and time can be defined for protection. But what happens in case of an oil accident? As of yet at the German North Sea Coast, only mechanical devices are used within the 20 m depth line for plans of response measures because of the high toxicity of chemical dispersants. However, later generations of these chemicals show a much lower toxicity so, that scenarios have to be examined as to whether or not the use of chemical dispersants should be applied. This evaluation belongs to:

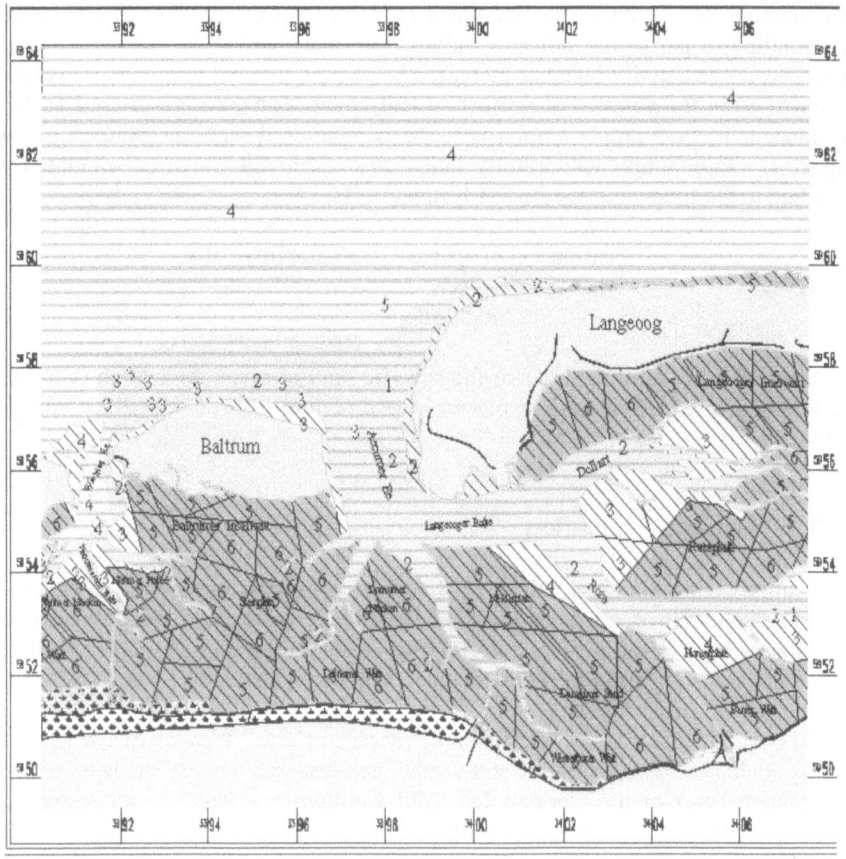

Fig. 7.4. Distribution of eu- and sublittoral areas with different sensitivity to oil pollution. The class-values for "benthos-sediment" and "birds" are indicated as numbers (1 – 6; the highest values 7 and 8 are not achieved)

7.4 Environmental (Ecological) Risk Assessment (ERA)

Whenever a decision-maker has to decide which kind of action can be applied, a risk evaluation occurs. Belluck et al. (1993) defined three classes for ecological risk assessment (scientific, regulatory, and planning) that lie along a continuum from most to least quantitative. Because costs (and usually time) increase with the level of scientific detail that can be obtained, the desire to improve the analysis must always be weighed against the cost of the additional information.

The behaviour of oil in water should be assessed before the questions "Will dispersants work effectively with a particular oil in the environment of interest?" and "What are the ecological consequences of dispersant use?" can be answered.

The main processes involved when oil (especially crude oil) is spilled on sea water summarised as "weathering" (Daling et al. 1990) are: spreading, evaporation, dissolution, formation of emulsions, dispersion in the water column, sedimentation and, biodegradation (Fig. 7.5).

Direction and speed of a driven oil slick depend on current conditions and about 3.5 % of wind velocity. Consequently, a drift model for coastal waters is a good tool for use as part of a conceptual model to predict the areas at risk. A spreading slick itself forms a large region of "sheen;" about 1 µm thick containing less than 5 % of the total oil volume. The majority of oil is bound to a much smaller area, with a thickness of several millimetres in case of a stable emulsion. Within the first few hours or days most crude oils will lose up to 40 % of their volume by evaporation. This process, driven by temperature and wind speed, reduces the portion of lighter components of the oil, leaving a smaller pollution volume with a higher viscosity and minor toxicity. This loss of oil components to the atmosphere is supplemented by a much smaller rate of dissolution. The amount of water soluble hydrocarbons around an oil slick is generally in the ppb range but remains toxic and bioavailable for marine organisms. On the other hand, the incorporation of water into the oil residue left by evaporation and dissolution leads to a large increase of pollutant volume, raising the viscosity once again. Very stable emulsions, formed by some oils, are resistant to chemical treatments or heating. Under rough sea conditions, low viscous oils disperse naturally into the water column to a large extent, forming droplets in a wide range of sizes. While larger oil droplets resurface, only the smaller (< 70 µ) are found in permanent dispersion. Clay and particles of similar size (1 – 100 µ diameter) and microscopic organisms interact with dispersed oil droplets by adsorption and ingestion. In waters of high turbidity, as for example in estuaries, the resulting oil-mineral complexes can reach high levels and obey the characteristic environmental processes of sedimentation and accumulation in areas of low hydraulic energy, like nearshore tidal flats and watersheds.

The final weathering-process of spilled oil is biodegradation. All but the most refractory components of a crude oil can be degraded by biological actions in the water column as well as in sediments. The rates depend on temperature and the availability of oxygen. They range from 1 – 50 mg/m³/day to years in very cold or anaerobic environments.

The advantages of using chemical dispersants are twofold. In the first place they reduce the pollutant volume on the water surface. Secondly they increase the

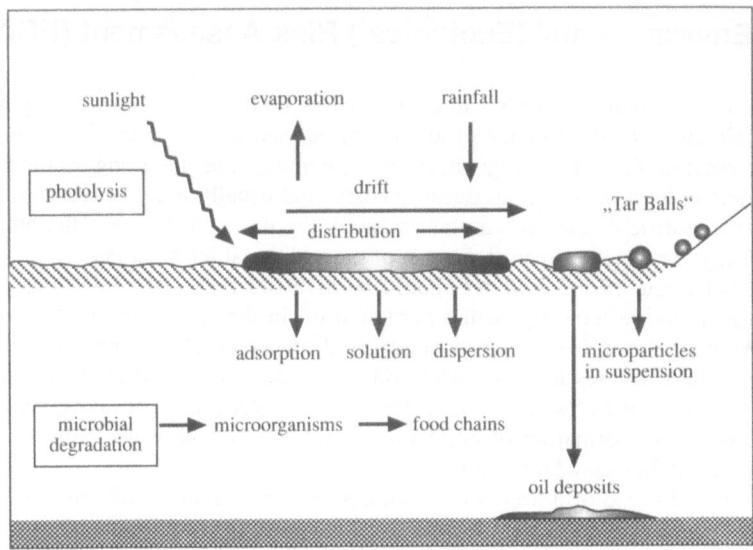

Fig. 7.5. Simplified pattern of oil spilled on water

rate of biodegradation processes by increasing the reactive surface of the oil. Their effectiveness depends mainly on the kind of oil, its state of weathering (viscosity and degree of water-in-oil emulsions) and on the hydraulic energy in the area of concern. Other factors of gradual influence are: salinity, turbidity and temperature.

There is no conclusive way to estimate their usefulness in a particular environment with different temporal conditions. Appropriate scenarios for individual cases have to be evaluated during a comparative analysis of the risks and benefits.

As a fundamental goal for oil spill response holds: to minimise the ecological impacts of a spill (Lindstedt-Siva 1991). The decision as to whether it is better to protect sensitive habitats rather than to optimise cleanup, needs a specific methodology to optimise all possibilities of response into an integrated programme. In this concern, a "Net Environmental Benefit Analysis" NEBA (Baker 1995) based on an ecological risk assessment approach can serve as part of integrated precaution measures.

The activities involved in the assessment can be summarised in three phases: problem formulation, analysis, and risk characterisation (US EPA 1992a).

Within these phases quantitative as well as qualitative data may be used depending on the state of knowledge about the systems involved. The uncertainty of data and methods has to be defined as far as possible before the resulting information can be incorporated into conceptual or mathematical models. Another key element as prerequisite to develop conceptual models is the identification of clear and consistent endpoints related to the protection of resources.

Applying these briefly depicted features to the environment at risk mentioned above, the Wadden Sea, we can establish the following characteristic aspects to define the limits for a selection of possible scenarios, which meet the basic question:

Is it possible to mitigate the damage of oil pollution by using chemical dispersants as (part of) the response measures (Hypothesis: If there is an increased degradation of oil and a decreased occurrence of oil slicks on the water surface, the ecological damage will be lower.)?

The coastal water of this area shows a wide range of salinity, turbidity and energy characteristics: With increasing distance from the coast, depending on the tide especially in estuarine areas, the salinity changes from less than 5 ‰ to more than 30 ‰. The gradient of turbidity is reverse but it is interrupted by high differences of about 80 to 1000 ppm. The heterogeneity of wave energy on a small scale (some 100 m) caused by changing wind conditions, water depth, and current speed in estuaries, tidal channels and creeks also decreases with distance from the coastline, while the heterogeneity and number of sensitive tidal and subtidal habitats increases. Briefly: the effectiveness of dispersants is greater the further offshore they are applied (wave energy and salinity); the danger to fundamental system functions is greater if they are used nearshore (high adsorption rate of oil droplets to particles leads to increased microbial degradation and is detrimental to oxygen content).

On the other hand, there are several nearshore phenomena of a high sensitivity to oil slicks: e.g. mussel beds, shell mounds, sea grass meadows, salt marshes and stocks of resting and moulting birds (van Bernem 1992). For most of these phenomena, an evaluation about the "good or bad" effects of chemical dispersant use also depends on the individual conditions of an accident; however, a special case exists with regard to moulting and resting birds. In particular, moulting bird stocks are clearly much more vulnerable to untreated oil slicks in comparison to chemical dispersions. During one to two months in the summer these birds are not able to fly because of changing their feathers. In distinct areas, these stocks can reach much more than 50,000 individuals; just swimming or drifting on the water they are helpless in the face of being contaminated by oil slick residues. Their stock sizes and population dynamics are very well known and steadily monitored so that the degree of uncertainty in estimating damages to the population level is comparably low. Although there may be no danger to the survival of their population and role in systems functions, nevertheless a rough decrease of their local stock size should be avoided for reasons of natural resource protection. Setting these conditions as endpoints for a special scenario of oil pollution response, the resulting simplified conceptual model would correspond to Fig. 7.6.

Thus, although there is only a small window left concerning chemical dispersants as part of response measures, it is possible to establish a decision tree for different scenarios of response measures and for different environmental conditions; if we set up distinct endpoints.

7.4.1 Environmental (Ecological) Impact Assessment (EIA)

Usually an EIA means a synopsis of assumptions, about which environmental effects are expected to happen following a planned intervention (coastal constructions, dredging measures or deposition of excavated sediments etc.). Corresponding to predictions made during an ERA for "unplanned" measures, an EIA can al-

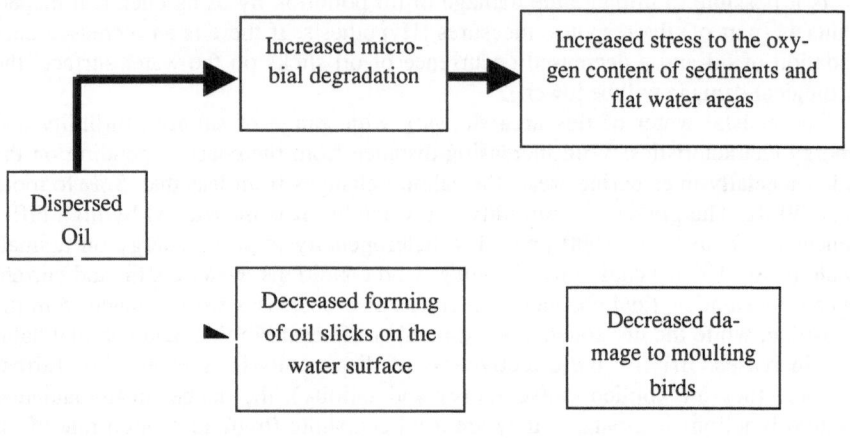

Fig. 7.6. "Good" and "bad" effects of dispersant use in oil spill response

so be done by an assortment of simplified conceptual models about the interrelationship of systems and disturbances. Because the interventions are planned, one is, in this case, well able to combine conceptions to predict effects with conceptions to verify the prognosis. The prerequisite is to fix the status quo ante conditions by a base line study and/or to identify reference areas respectively. According to the "decision endpoints" mentioned above, it is also necessary to set "measurement endpoints" for a resulting assessment, because it is impossible to monitor all conceivable environmental effects of an intervention. The features defining these measurement endpoints depend on the environmental attributes and functions considered suitable in indicating possible effects, as well as on the reliability and accuracy of methods to detect significant differences.

These conditions are described as follows, using the landfall of a gas pipeline at the Wadden Sea coast as an example.

7.4.2 The Construction Measures

The pipeline ($\varnothing \sim 1.2$ m), called "Europipe," approximately 670 km long, was constructed to connect the gas fields of the Norwegian North Sea with the city of

Emden in Northwest Germany. It was laid, uncovered, offshore on the seabed as far as the 15 m depth-line. From there it proceeded landwards in a ditch, about 12 km long, which was excavated to a depth of 5 m, (15 m floor-width, about 100 m surface-width), by seagoing dredging-vessels. Around the reef-bow of the tidal channel between the barrier islands Baltrum and Langeoog it was found necessary to dig an auxiliary channel for the pipe-laying vessel. An anchor-corridor was prepared for mooring the vessel during the excavation. Landward of the exposed part of the tidal channel, in the sheltered "Accumersieler Balje," up to a tie-in-chamber the pipeline was layed using smaller equipment. Following a "decision endpoint" of national park authorities not to excavate any sediments on tidal flats, these areas were crossed by a tunnel ($\varnothing \sim 3.8$ m, about 2 km long) at a depth of about 10m to the pit-opening behind the dike (Fig. 7.7).

During the excavation (including maintenance-dredging and substitute- water-way) about 3.6×10^6 m^3 of sediment arose, which were temporarily stored in the

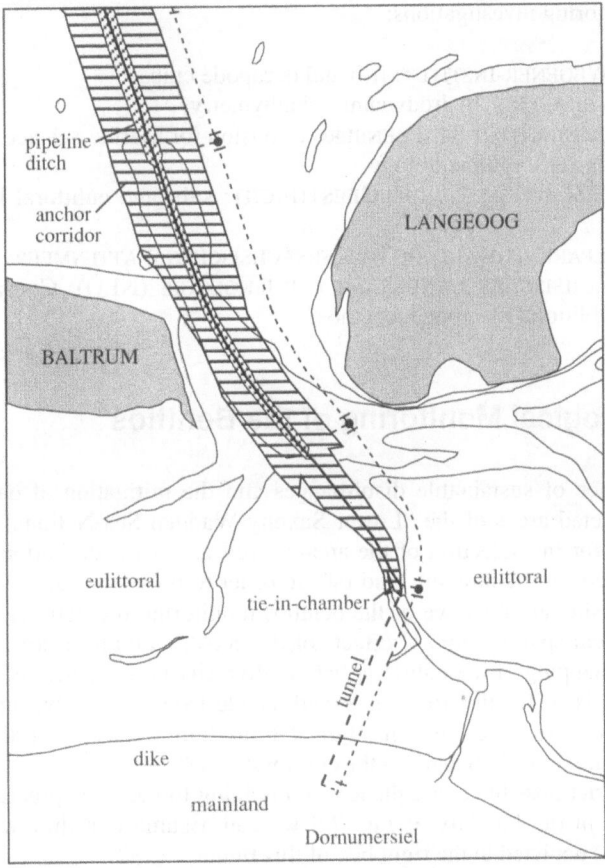

Fig. 7.7. Course of the laying-ditch through the tidal channel and of the adjacent tunnel-line. The areas covered by the ditch including the anchor-corridor are hatched

north of Langeoog and to a minor degree in the tidal channel (about 60,000 m^3), after clay and peat had been divided off.

Approximately 3.8 x 10^6 m^3 of sediment were needed to refill the ditch, so that the total quantity of sediment moved was about 7.5 x 10^6 m^3.

7.4.3   The Integrated Ecological Monitoring-Investigations

In accordance with the regional planning assessment as a completion of the regional development procedures as well as the decision of the official planning approval proceedings, following the mining law, an ecological monitoring was undertaken, before, during and after the construction measures. The ecological monitoring was part of the environmental protection programme and was established by STATOIL on the basis of the British Standard BS 7750 "Specification for Environmental Management Systems" on the background of the "European-Ecological-Audit-Regulation." The following institutions participated in the ecological monitoring investigations:

ALFRED-WEGENER-INSTITUT: fish and decapode crabs.
DELFT HYDRAULICS: hydrodynamics, bathymetry.
FORSCHUNGSINSTITUT SENCKENBERG (Division for Marine Science, Wilhelmshaven): sedimentology.
GKSS-FORSCHUNGSZENTRUM GEESTHACHT: sub- and eulittoral benthic communities.
NATIONALPARKVERWALTUNG NIEDERSÄCHSISCHES WATTENMEER: avifauna
NIEDERSÄCHSISCHES LANDESAMT FÜR ÖKOLOGIE (NLÖ), Coastal Research Centre: sublittoral macroendobenthos.

7.5 Ecological Monitoring of the Benthos

The avoidance of sustainable disturbances and the mitigation of damage to the biota in affected areas of the "Lower Saxony-Wadden Sea-National Park" was a prerequisite for the selection of the area as well as for the definition of technical measures used for the pipeline land fall. In order to prove the success or failure of this prerequisite, an objective of the benthos monitoring program was to scientifically document spatio-temporal effects of the construction measures. Except for a large scale mapping of the eulittoral habitat diversity (van Bernem et al. 1992), no basic data existed for the area concerned. Aside from this study, further surveys were based on the underlying question: Do dredging measures in the sublittoral zone cause measurable changes in the eulittoral zone?

If we restrict possible consequences of dredging to the basic processes depicted from 1 to 5 in the left box of Fig. 7.8 we can assume that they will cause the events (1 – 5) depicted in the right box of this figure.

The next conception concerns the problem of how these events can be assessed with sufficient reliability. In other words, we have to see whether there are suitab-

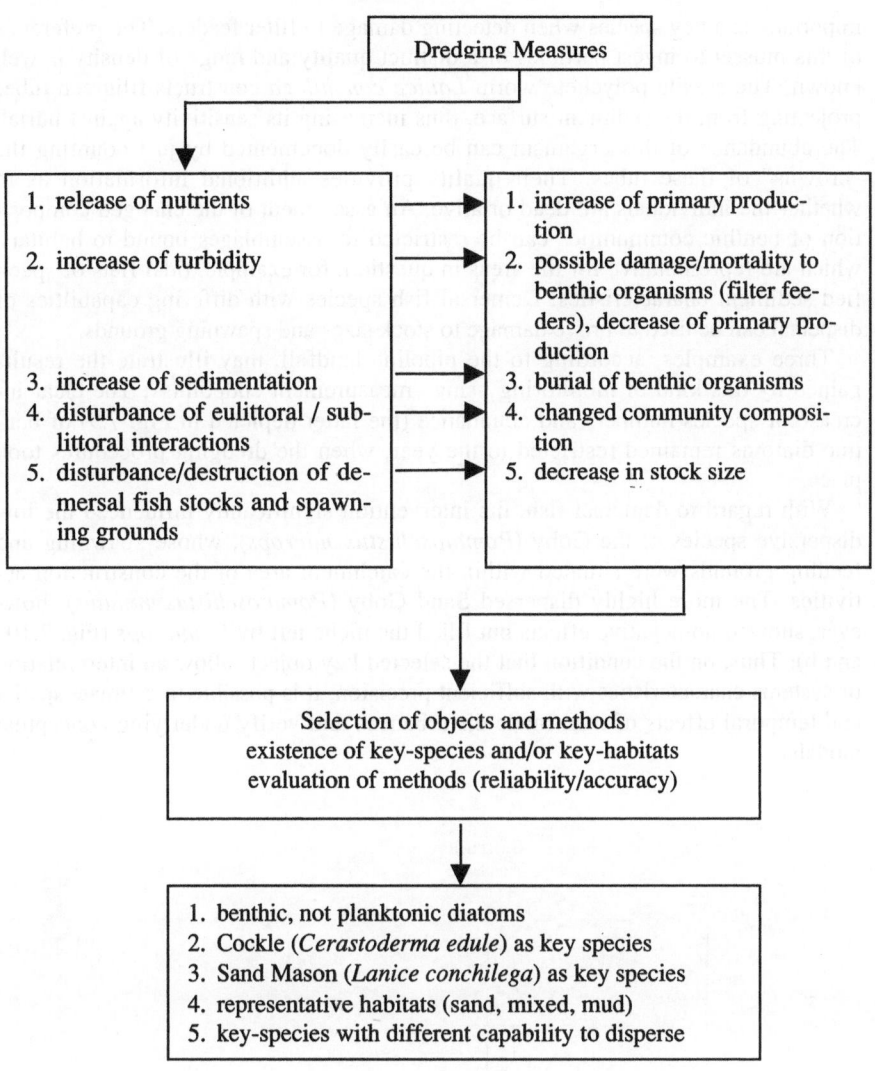

Fig. 7.8. Hypothesised effects of dredging measures and selection of objects and methods for monitoring

le key objects (species, processes or habitats) that can be verified with particular methods.

Primary production, for example, can be measured by several methods. If we decide to use organismic procedures in order to achieve additional information about the community conditions (according to hypotheses 1 and 4) we should use benthic rather than planktonic diatoms, because they can more easily be sampled representatively. For the same reason, the Cockle (*Cerastoderma edule*) may be important as a key species when detecting damage to filter feeders. The preference of this mussel to ingest particles of a distinct quality and range of density is well known. The sessile polychete worm *Lanice conchilega* constructs filigreen tubes projecting from the sediment surface, thus increasing its sensitivity against burial. The abundance of this organism can be easily documented by just counting the "crowns" of these tubes. Their quality provides additional information as to whether the individuals are dead or alive. An assessment of the changed composition of benthic communities can be restricted to assemblages bound to habitats, which are representative for the areas in question, for example, tidal flats of specified sediment characteristics. Demersal fish species with differing capabilties of dispersal can be used to prove damage to stock sizes and spawning grounds.

Three examples, according to the pipeline-landfall, may illustrate the results gained by this kind of monitoring using "measurement endpoints": The clear increase of species numbers and abundance (the latter depicted in Fig. 7.9) of benthic diatoms remained restricted to the year, when the dredging procedures took place.

With regard to demersal fish, the intervention significantly influenced the low dispersive species of the Goby (*Pomatoschistus microps*), whose spawning and feeding grounds were situated within the catchment area of the construction activities. The more highly dispersed Sand Goby (*Pomatoschistus minutus*), however, showed no negative effects but filled the niche left by *P. microps* (Fig. 7.10a and b). Thus, on the condition that the selected key objects allow an interpretation of systems characteristics with sufficient precision, it is possible to estimate spatial and temporal effects of man-made disturbances and verify underlying conceptual models.

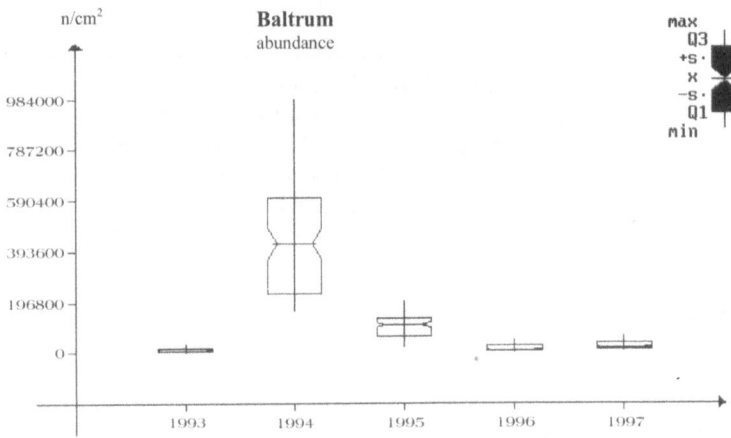

Fig. 7.9. Abundance (150 subsamples/year) at a location leeward side of the island of Baltrum. The values for 1994 are statistically different compared to the other years (H-Test, Kruskal and Wallis, $P \leq 0{,}05$)

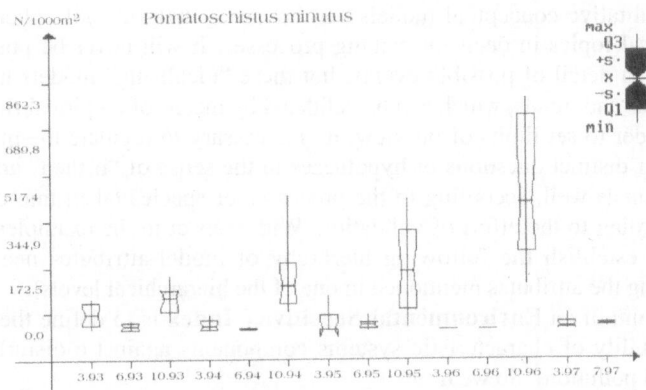

Fig. 7.10a. Group mean values for the abundance on 8 transects (10 hauls each) per season (For further explanations, see text)

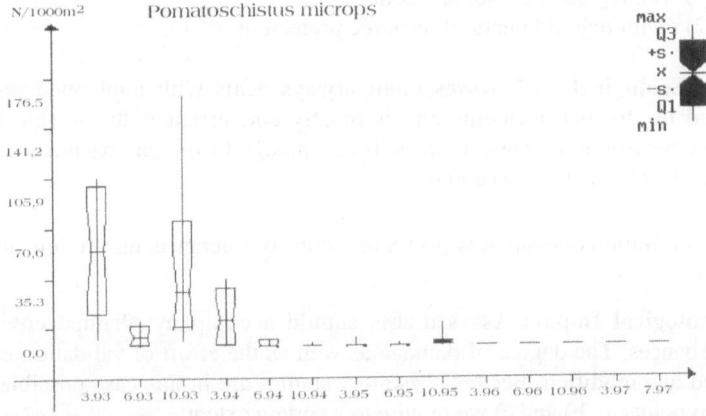

Fig. 7.10b. Group mean values for the abundance on 8 transects (10 hauls each) per season (For further explanations, see text)

7.6 Conclusions

The high variability of complex biological interactions causes large statistical and systematical errors, which make it impossible to realistically reflect nature using complex simulation models. An adequate correspondence with reality, however, is a prerequisite, when considering the tolerance of ecosystems to man-made disturbances and countermeasures. If we restrict the underlying problems to soundly investigated systems, well-known key parameters, and short term-effects, simplified qualitative conceptual models can be a very helpful tool, when considering ecological topics in decision-making processes. It will never be possible to consider each detail of possible events, but these "idealising" models help to clarify principles and trends, which can be validated by means of empirical findings.

In order to set limits of purview, it is necessary to regulate the models following clear distinct questions or hypotheses in the sense of "if/then" and set defined endpoints as well, according to the processes or species taken into consideration, as according to the effort of validation. With respect to the examples represented, we can establish the following hierarchy of model-attributes needed (without excluding the attributes mentioned in one of the hierarchical levels):

The aim of an **Environmental Sensitivity Index** is to define the fundamental vulnerability of characteristic systems components against a disturbance (in our case: oil pollution). So we need:

a) conceptual models with regard to basic functions/processes/key species of the systems involved, and

b) conceptual models of the possible interactions between systems and disturbance
   b-1) with regard to ecological conditions
   b-2) with regard to natural resource protection

An **Ecological Risk Assessment** always deals with unplanned disturbances caused by distinct accidents and is mostly concurrent with the selection of response measures. Because there will be damage to the environment in any case, we need additionally to a) and b):

c) the definition of endpoints (to fix the limits of tolerance, uncertainty and effort)

Ecological Impact Assessments should accompany planned environmental disturbances. The degree of damage as well as the effort of validation can be controlled and modified. Because base-line studies are in this case possible, additionally to points a), b) and c) we require to a certain extent:

d) conceptual models for monitoring, according to the measures planned.

Chapter 8
Models in the Mechanics of Materials

by Wolfgang Brocks

Abstract

A modern concept of materials characterisation has to be based on three "pillars" of modeling, namely testing, constitutive theories and numerical simulations. The contribution aims to describe the intellectual process of abstraction which was necessary in order to understand, describe, calculate, and predict the mechanical behaviour of materials and structures. The presentation is not meant for specialists in either of these subjects but wants to make the basic ideas clear to people generally interested in natural sciences.

8.1 Introduction

Modeling has become an important and fashionable issue. Every serious research project will claim modeling activities to increase the chances of being awarded grants. Modern technology and product development have detected the saving effects of modeling: "The development and manufacture of advanced products, such as cars, trucks and aircrafts require very heavy investments. Experience has shown that a large portion of the total life cycle cost – as much as 70 – 80 percent – is already committed in the early stages of the design. It is important to realise that the best chance to influence life cycle costs occurs during the early, conceptual phase of the design process. Improvements in efficiency and quality during this phase should enable us to obtain the right solutions and make the right decisions from the beginning. This requires good design, analysis and synthesis methods and tools, as well as good simulation techniques, including computational prototyping and digital mock-ups" (Fredriksson and Sjöström 1997).

"Modeling," however, is an ambiguous term and needs further explanation and a more precise definition. The common understanding of a model is manifold (Collins 1998):

1. a three-dimensional representation, usually on a smaller scale, of a device or structure: *an architect's model of the proposed new housing estate.*
2. an example or pattern that people might want to follow: her success makes her an excellent role model for other young Black women.

3. an outstanding example of its kind: *the report is a model of clarity.*
4. a person who poses for a sculptor, painter, or photographer.
5. a person who wears clothes to display them to prospective buyers; a manne-
   quin.
6. a design or style of a particular product: the cheapest model of this car has a
   1300cc engine.
7. a theoretical description of the way a system or process works: *a computer
   model of the British economy.*

For natural and engineering sciences, we shall generally adopt items #1 and #7
as definitions. In a broad sense, every scientific activity might be looked at as
"modeling", since dealing with a complex reality always requires reduction and
idealisation of problems. Thus, modeling may be understood as novel only in the
sense of "computational simulation of reality," which is the underlying compre-
hension in the quotation "simulation techniques including computational prototyping" given above (Fredriksson and Sjöström 1997). The intention of the present
contribution is to show that, at least in engineering sciences, "modeling" has to
combine and integrate computational and experimental efforts, in order to proceed
to an understanding of the physical phenomena, which allows for realistic predictions of the performance, availability and safety of technical products and systems.

8.2 Modeling in the Mechanics of Materials

8.2.1 Testing

Materials testing has a long tradition and is based on the desire of scientists to
measure the mechanical properties of materials and the need of design engineers
to improve the performance and safety of buildings, bridges, and machines. A
stimulating view into its history can be found in Ruske (1971), where the author
found all the historic pictures that are reproduced later in this section. Mechanical
sciences started with Galileo Galilei (1564 – 1642). He did not only promote Co-
pernicus' concept of a heliocentric planetory system, but studied the laws of fal-
ling bodies and strength of materials both theoretically and experimentally (Galilei
1638). An actual engineering problem was the dependence of the strength of a
bending bar on its cross sectional dimensions for which Galilei designed an experiment shown in Fig 8.1. The test configuration reduces the complex problem of
structural bars, e.g. in housing, to a cantilever beam under a single load at its end.
He found that the "bending resistance" was proportional to the width, b, and the
square of the height, h, of the bar.

Expressing this result in modern mathematical terms, we can derive today that
the section modulus is $W = 1/6bh^2$. Neglecting the weight of the bar we find the
bending moment becomes $M = G\,l$, where G is the applied weight E at the end of
the bar of length l. Finally, the maximum tensile stress occuring in point A results
from $\sigma_{max} = 6\ Gl/bh^2$, but these mathematical formulas and a general theory of
bending did not exist in Galilei's time. They were developed about one and a half

Fig. 8.1. Galilei 's experiment on the bending problem of a bar (Galilei 1638)

century later by mathematicians like Jakob Bernoulli (1655 – 1705) and engineers like Ch. A. Coulomb (1736 – 1806) and L. M. H. Navier (1785 – 1836), who introduced new concepts and abstract ideas like "bending moments," "stresses," and "strains," which allow one to relate "bending strength" with "tensile strength." Galilei also did not consider the deformation of the bar, either, as the law of elasticity, later found by R. Hooke (1635 – 1703), was unknown. The "section modulus," W, is a purely geometrical quantity which is determined by the shape and dimensions of the cross section. Thus, Galilei's experiment actually did not reveal material properties.

The obvious question that arises from any experiment is: What can we learn from it? – or more precisely: how does this configuration compare to the "real" situation? For instance, can we take the fracture load obtained in the above test to design the supporting beams in a building? Finally, we reach the fundamental and still present-day problem of materials testing: is the test data measured from a specimen transferable to a real and large scale structure? Specimens used in materials testing are models in the sense of "a three-dimensional representation, usually on a smaller scale, of a structure" (see Sect. 8.1). In addition, they are of a simpler geometry and under simpler loading conditions. The issue whether the information

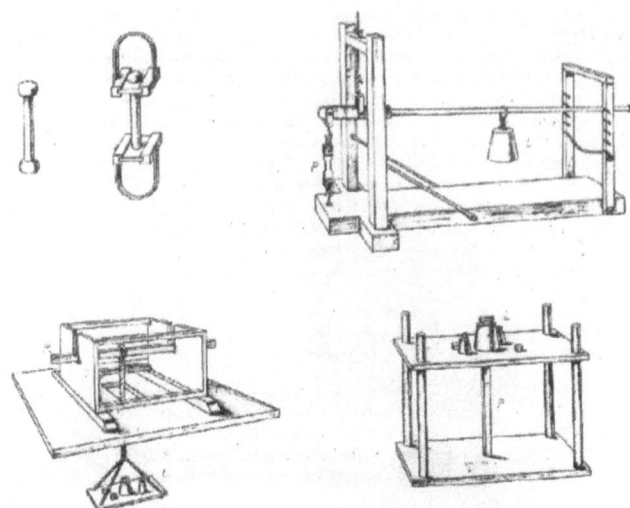

Fig. 8.2. Specimens and set-ups for tensile (upper), bend (lower left), and compression test (lower right) designed by Pieter van Musschenbroek (1729) (picture taken from Ruske 1971)

from a (simplified) model may or may not be transferred to (complex) reality is still controversial in many cases.

A deeper understanding of Galilei's bending problem would have required a theory that did not exist in the 17th century. Nevertheless, engineers wanted to design structures and get information on the mechanical behaviour of different

Fig. 8.3. Tensile testing machine by Mohr and Federhaff machine Company, Mannheim 1870, with hydraulic drive and balance for measuring force (picture taken from Ruske 1971)

materials. Hence, they had to develop special test set-ups for various loading conditions like tension, compression, bending, buckling, etc. (see Fig 8.2).

The cylindrical test rod shown in the upper left of Fig 8.2 is not much different from current standard tensile specimens. The testing machines, however, became much more "sophisticated" during the following years (see Fig. 8.3).

With expanding technology, other material properties became relevant, not only under static loading but also under impact or oscillating stresses. Scientists and engineers had found that the "ductility" of a metallic material was an important property, which influences the safety margins of a structure or plant. This ductility depends on the deformation rate and the service temperature. It was also found that materials may "fatigue" and fail under operating stresses much less than their static strength, if the loads are oscillating and structural parts are subject to a high number of load cycles. Thus, new test facilities were designed and built like the pendulum impact testing machine by the French metallurgist G. Charpy in 1901 (Charpy 1901), to measure the energy necessary to fracture a notched bar, Fig. 8.4, and the testing machine for alternating tensile loads by the German materials scientist A. Wöhler (1819 – 1914), to determine the fatigue strength of a material (see Fig 8.5).

Both the Charpy impact test and Wöhler's fatigue test are still in use today to qualify structural materials, although the instrumentation and control systems have become more and more sophisticated over the years. However, the fundamental problem of materials testing, i.e., how much these tests tell us about inherent material properties, has still remained controversial. Separating "material properties" from "structural properties" is an intellectual process of abstracting which is typical for "modeling."

Wöhler's fatigue testing machine in Fig. 8.5 is an impressive example that ad-

Fig. 8.4. Pendulum impact testing machine by Charpy (picture taken from Ruske 1971)

Fig. 8.5. Testing machine for alternating tensile loads of four specimens by Wöhler, Berlin 1860 (picture taken from Ruske 1971)

Fig. 8.6. Triaxial testing machine by Instron-Schenck, force or displacement controlled tensile or compressive loading in three axes; GKSS Research Centre, Geesthacht, 1998

vanced testing may be done on "small scale" specimens but requires a comparably huge testing facility so that the definition given above of a model as a representation "usually on a smaller scale" may appear obscure. However, it is well known from elementary physics that small particles may require even bigger test setups. A quite sophisticated modern testing machine, which allows for arbitrary loading histories in tension and compression in all three space directions to investigate the influence of multiaxial stress states and non-proportional loading on the deformation and damage evolution, is shown in Fig. 8.6.

Summarising this historical excursion, we have to keep in mind that right from the start materials testing means modeling with its two important characteristics of

- reduction of the complexity of a problem (geometry, loading),
- idealisation of the process (static, impact, alternating loads).

These tests on comparatively simple specimens are performed in order to obtain information on the materials strength and toughness and to conclude to the mechanical behaviour and performance of complicated structural geometries under different kinds of loading and various loading histories. This understanding requires a theory, which has been developed in the late 19th and early 20th century and been continously improved ever since.

8.2.2 The Theory of Continuum Mechanics

Describing the full history of engineering mechanics would fill books. The following section intends to impart the fundamental concepts of the modern theory of continous solid media, which forms the basis of current developments in the mechanics of materials. The presentation follows the description in (Truesdell and Noll 1965) but can be found in any modern textbook of continuum mechanics as well (e.g., Eringen 1967; Hodge 1970; Gurtin 1981; Lai et al. 1993).

Continuum mechanics is a phenomenological theory, which means it is based on observed phenomena and establishes mathematical models for the mechanical behaviour of matter. It is a field theory, as it deals with physical quantities like deformations, stresses etc., which are spatially distributed. The behaviour of matter is determined by interactions of atoms and molecules, as everybody has learned in physics. However, a modeling for engineering purposes cannot be done on this level. The discretely structured matter is hence represented by a phenomenological model, i.e., the continuum; this is done by averaging its properties in space.

For describing the mechanical behaviour of heterogeneous materials with strong local gradients of the microstructure, this approach is in general not sufficient. For this purpose, micromechanical constitutive models are currently being developed and applied. Micromechanical modeling takes place on a "meso level" between a full representation of the microstructure and a phenomenological approach. This will be further illustrated later.

The equations of continuum mechanics can be separated into two classes:

- material independent descriptions and principles of kinematics and kinetics, and
- material dependent "constitutive" equations.

Kinematics defines and describes terms and geometrical quantities like bodies, configurations, motions, deformations, whereas kinetics deals with physical entities like forces as "external actions" and stresses as "internal reactions." Balance equations are formulated for the physical quantities, namely conservation of matter and balance of energy, work, or power, i.e., the first law of thermodynamics. Alternatively to the first principle, conservation of momentum and moment of momentum are postulated for static problems or equations of motion for dynamic problems. The second principle of thermodynamics, i.e., the entropy inequality, is also material independent and gives the frame for all thermodynamically consistent constitutive equations. However, it does not allow the derivative of a general law for the mechanical behaviour of materials. The combination of kinematic equations, physical principles, constitutive laws, initial conditions (in time), and boundary conditions (in space) leads to the formulation of a set of commonly highly nonlinear differential equations, a so-called initial boundary value problem. At this point, modeling is finished and followed by mathematical efforts to solve the equations.

Just to help understanding of the mathematical concept of continuum mechanics and the mechanical terms used in the following, some definitions are given:

- A body, $\mathcal{B}$ is a three-dimensional differentiable manifold, the elements of which are called particles $x \in \mathcal{B}$. A body is endowed with a non-negative scalar measure, m, which is called the mass of the body.
- A configuration, χ of a body $\mathcal{B}$, is a smooth homeomorphism of $\mathcal{B}$ onto a region, $\mathcal{B} = \{x = \chi(x) \mid x \in \mathcal{B}\} \subset \mathbb{E}^3$, of the three-dimensional Euklidean space $\mathbb{E}^3$, called the region occupied by the body $\mathcal{B}$ in the configuration χ.
- A motion of a body $\mathcal{B}$ is a one-parameter family, ${}^t\chi$, of configurations, the real parameter t being the time. The point $x = {}^t\chi(x) = \chi(x, t)$ is the place occupied by the particle x at time t.

The definition of a body, given here, is something completely abstract. Especially, a body should not be mixed up with the region, which it occupies at some reference time. The mass of a body is a fundamental physical property and hence is assigned *a priori* as a part of the specification of the body. The relation between the mass $m(\mathcal{P})$ of every measurable part $\mathcal{P}$ of $\mathcal{B}$ and the volume $V(\mathcal{P})$ of the region $\chi(\mathcal{P})$ in Euklidean space, which the part $\mathcal{P}$ happens to occupy in the configuration χ, is given by the mass density, ρ:

$$m(\mathcal{P}) = \int_{\mathcal{P}} \rho(x)\, dV \tag{8.1}$$

Conservation of mass states that $m(\mathcal{P},t) = m(\mathcal{P})$ or $\dot{m} = 0$.

Any general motion of a body includes a deformation that can be described as follows. A particle, x, which is identified by its place $^0x = {}^0\chi\,(x)$ in a reference configuration at time $t = 0$, occupies the place

$$^t x = {}^t\chi(x) = \chi\left({}^0\chi^{-1}({}^0x),t\right) = \phi\left({}^0x,t\right) = {}^t\phi\left({}^0x\right) = {}^0x + {}_0^t u \qquad (8.2)$$

at the time t. The mapping $^t\phi$ is called deformation and relates the reference configuration to the present configuration, and $_0^t u$ is the displacement of the particle. Observing an infinitesimal vicinity of the particle in the course of motion,

$$^t x + d^t x = {}^t\phi\left({}^0x + d^0x\right) \approx {}^t\phi\left({}^0x\right) + \frac{\partial\, {}^t\phi}{\partial\, {}^0x}\cdot d^0x, \qquad (8.3)$$

leads to the definition of the deformation gradient,

$$_0^t F = \frac{\partial\, {}^t f}{\partial\, {}^0x} = {}^0\text{grad}\,{}_0^t u + I, \qquad (8.4)$$

with I being the second order unit tensor. Various tensor-valued measures of the local deformation, commonly called strains, can be derived from $_0^t F$. The tensor of linear, engineering strains

$$_0^t E = \tfrac{1}{2}\left[\,{}^0\text{grad}\,{}_0^t u + \left({}^0\text{grad}\,{}_0^t u\right)^T\right] \qquad (8.5)$$

is one of them, but several others can be defined. No further details on deformation measures, their physical meaning and proper use will be given here (see Eringen 1967; Hodge 1970, Gurtin 1981, Lai et al. 1993).

The motion (and deformation) of a body is caused by external actions, called forces, on the body $\mathcal{B}$. The concept of forces does also describe the interaction between the different parts $\mathcal{P}$ of the body. Forces are mathematically characterised by the vector fields of

- external body forces, $b\,(x,t)$, defined for any $x \in \mathcal{B}$ and
- the contact forces or tractions $t\,(x,t;\,\mathcal{P})$, defined for any x on the boundary $\partial\mathcal{P} \subset \mathcal{P} \subseteq \mathcal{B}$.

The total resultant force, $f\,(\mathcal{P})$, exerted on any part $\mathcal{P} \subseteq \mathcal{B}$ is defined as the sum of the resultant body force and the resultant contact force

$$f\,(\mathcal{P}) = \int_{\mathcal{P}} b\rho\,dV + \int_{\partial\mathcal{P}} t\,(x;\mathcal{P})\,dA. \qquad (8.6)$$

There is a vector-valued function $t(x,n)$, called the stress-vector, defined for all points x and unit vectors n, such that the tractions acting on $\mathcal{P}$ are given by

$$t\ (x; \mathcal{P}) = t\ (x, n),$$ (8.7)

where n is the exterior unit normal vector at the point x on the boundary of $\mathcal{P}$. Under suitable continuity conditions, there also exists a stress-tensor field $T\ (x)$, such that

$$t\ (x, n) = T(x) \cdot n$$ (8.8)

The fundamental principles of momentum and moment of momentum are then equivalent to Cauchy's laws of motion

$$
\begin{aligned}
div\ T + \rho b &= \rho \ddot{x} \\
T &= T^T
\end{aligned}
$$ (8.9)

For static problems, we have $\ddot{x} = 0$, and Eq. 8.9 is the equilibrium condition. Alternatively to the formulation of differential Eq. 8.9 governing the behaviour of an infinitesimal region, a variational or extremum principle valid over the whole region occupied by the body can be postulated, and the correct solution of the problem is the one minimising some functional Π which is defined by integration of the unknown quantities over the whole domain. A common principle used in finite element formulations, e.g., (Abaqus 1998) is the principle of virtual work

$$\delta\Pi = -\int_{\mathcal{B}} T \cdot \cdot \delta(\mathrm{grad}\,v)\,dV + \int_{\partial \mathcal{B}} t \cdot \delta v\,dA + \int_{\mathcal{B}} \rho b \cdot \delta v\,dV = 0$$ (8.10)

where the "virtual" velocity field, δv, is completely arbitrary except that it must obey any prescribed constraints and have sufficient continuity. Equation 8.10 follows directly from Eq. 8.9 for $\ddot{x} = 0$ with Gauss' theorem and by integration over $\mathcal{B}$. It is often called the "weak form" of Cauchy's equilibrium equations, as it establishes equilibrium by a single scalar equation over the entire body. But as δv is arbitrary, and Eq. 8.10 has to hold for any part $\mathcal{P} \subseteq \mathcal{B}$ of the body; the two approaches (8.9) and (8.10) are mathematically equivalent; an exact solution of the one being the solution of the other.

As stated before, all the equations above represent physical principles, which are material independent. They are, of course, not sufficient to determine the motion and deformation of a body subject to forces. Additional equations describing the specific relations for a given material among stresses, T, and deformations measured in terms of ${}_0^t F$, the so-called "constitutive equations" are needed. This is the main business of "Constitutive modeling" in the mechanics of materials. These relations are in general nonlinear and may even include the whole deformation history of a body. There is no way of deducing them from general principles; they must be established individually for different classes of materials and phenomena. They have to fulfill some fundamental principles, however, namely

- the principle of determinism stating that the stress in a body is determined by the history of the motion of that body,

- the principle of local action stating that in determining the stress at a given particle **x**, the motion outside an arbitrary neighbourhood of **x** may be disregarded,
- the principle of frame-indifference stating that constitutive equations must be invariant under changes of the frame of reference, and
- the second law of thermodynamics stating that the entropy may not decrease in a physically admissable process.

Of course, constitutive equations should ensure that physically reasonable problems should have physically reasonable solutions, but unfortunately this "engineering requirement" cannot be put into a precise form at the present time.

It must be acknowledged that a full treatment of constitutive modeling can fill books, e.g. (Lemaitre et al. 1990; François et al. 1998), and any attempt to give an overview would exceed the limited number of pages available. The presentation of the basic equations of continuum mechanics given above is intended to show the considerable work of abstraction, which has to be carried out to draw the appropriate conclusions from the observed phenomena of the mechanical behaviour of materials. The important essence of this theory is that it allows the separation of "material properties" from effects of the geometry. Just performing tests and measuring forces and deformations are not sufficient for characterising a material and determining material parameters. A material characterisation by certain parameters is possible and meaningful only in the context of "constitutive models" which describe the mechanical behaviour. These models have to be established for general, i.e., three-dimensional situations, in order to overcome the problem of transferability of the characteristic parameters which are determined on test specimens under, commonly, uniaxial loading and uniform stress and strain fields. This ideal conception of material characterisation, however, is not yet fully realised, as it would need an interdisciplinary cooperation of experts in materials science, continuum mechanics, and numerical analysis at least, but its necessity is obvious.

8.2.3 Numerical Analysis

Analytical solutions of the equations describing the structural and material behaviour can be found in the theory of elasticity in a number of problems as the equations are linear, provided that the deformations are small. In general, the initial boundary value problem is highly nonlinear, due to large deformations or nonlinear constitutive equations. Despite a few analytical solutions in the theory of plasticity, numerical methods are required to solve the problems. The two formulations of problems in continuum mechanics mentioned above, namely as an initial boundary value problem or as a variational problem, call for different numerical solution procedures. Finite difference techniques approach the solutions of differential equations directly by approximating to those in a discrete manner. Other numerical techniques like Ritz's method and its variant, the finite element method, deal with an approximate minimisation of a potential, Π, or the principle of virtual work, Eq. 8.10, respectively.

The finite element (FE) method (Zienkiewicz 1971; Bathe 1996), which will be shortly outlined, has not only become the most versatile tool for structural analyses, but also a general method of wide applicability to engineering and physical science problems. It is essentially an approach through which a continuum with infinite degrees of freedom is approximated by an assemblage of subregions, or "elements," of a specified, finite number of unknowns (see Fig 8.7). Each such element interconnects with others at a discrete number of nodal points. If the force-displacement relationships for the individual elements are known, it is possible to calculate the variational functional, $\delta\Pi$, by a summation of the contributions of all elements; and from this, the deformation of the assembled structure under given external loads is derived. The method applies to many problems of non-structural type, as well, e.g., in problems of interconnected electrical circuits or fluid-carrying pipes. The general procedure of assembly and solution follows a pattern for which the structural analogy provides a convenient basis.

The modeling and computational procedure follows the general pattern:

- The continuum is separated by imaginary lines or surfaces into a number of finite elements (see Fig 8.7).
- The elements are assumed to be interconnected at a discrete number of nodal points situated on their boundaries. The displacements of these nodal points are the basic unknown parameters of the problem.
- A set of functions is chosen uniquely to define the state of displacement, u, within each element in terms of its nodal displacements.
- The displacement functions uniquely define the state of deformation, i.e., some strain tensor derived from $_0^t F$, Eq. 8.4, within an element in terms of the nodal displacements. These strains, together with the constitutive properties of the material, determine the state of stress, T, throughout the element and, by Eq. 8.8, also on its boundaries.
- A system of forces concentrated at the nodes, Fig 8.7, equilibrating the boundary stresses, t, is determined, resulting in a force-displacement or "stiffness" relationship for each element.
- Nodal displacements, nodal forces and element stiffnesses are assembled according to the conditions of connectivity for all elements to compose the system of equations, ensuring the conditions of compatibility and equilibrium throughout.
- Any system of nodal displacements listed for the whole structure in which all the elements participate automatically satisfies the condition of compatibility. As the equilibrium condition has already been satisfied within each element, all that is necessary is to establish equilibrium at the nodes of the structure.
- The resulting equations governing the mechanical behaviour of the entire structure contain the nodal displacements as unknowns. A solution of this system of equations provides an approximate solution of the fields of displacements, strains and stresses throughout the domain of the body.

The discretisation process of dividing a given structure into elements, specifying the nodal connections and prescribing displacement constraints and external forces is called "meshing" and has to be done by the user, either manually or computer aided. The choice of functions defining the displacement field within each element is done by choosing an "element type" offered in the element library of every commercial FE programme. Additionally, the user has to choose a constitutive model from the materials library of the programme – or provide his own model. The rest is then done by the computer programme.

For small, linear deformations and linear constitutive equations as in elasticity, the system of equations governing the mechanical behaviour of the structure is linear and can be solved easily. But in general, the kinematics of the problem is nonlinear, and the material behaviour is nonlinear and history dependent. In this case, the structural analysis will require iteration procedures for establishing equilibrium and a step-wise, incremental tracking of the loading history.

Commercial FE programmes offer a more or less large library of elements and material models. However, they are limited to a certain "state of the art," e.g., elasticity, metal plasticity, viscoplasticity, etc. New materials as well as the application of classical materials under increasingly extreme mechanical and loading conditions will possibly not be adequately described by the available material routines. It is the user's task, then, to develop appropriate, improved material models and the respective numerical algorithms to be implemented in the FE code.

Two examples shall illustrate the application of FE analyses in materials science.

8.3 Examples

8.3.1 The Tensile Test

The tensile test is an old and simple, nevertheless still useful method for determining mechanical properties of, mostly, metallic materials (see Figs. 8.2 and 8.3). A cylindrical bar of standardised size is subject to a tensile force and its elongation is measured. The information which can be taken from such a test is manifold:

- The ratio of stresses and strains, as long as their relation is linear, defines Young's modulus of elasticity.
- The beginning of a non-linear shape of the load-displacement curve indicates the onset of irreversible, "plastic" deformation and defines the yield strength.
- The point of maximum applied force defines the "ultimate tensile strength;" it is followed by a non-uniform elongation and necking of the bar.
- Elongation and reduction of area at fracture of the bar give a technological characterisation of the ductility of the material.

Only the items #1 and #2 can be considered as material properties; the two others, namely the onset of necking and the reduction of area at fracture are affected by the specific geometry and loading condition of a tensile bar. The underlying assumption of a uniaxial stress state and uniform stress and strain fields is violated beyond uniform elongation.

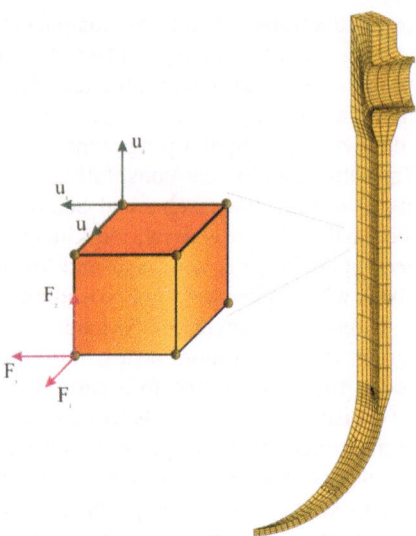

Fig. 8.7. FE model of a sector of a cylindrical pressure vessel accounting for its symmetry (right) and a single three-dimensional solid element with nodal displacements and forces (left)

Computational methods and modern constitutive models have extended the expressiveness of tensile testing considerably. With a continuum theory as outlined above, three-dimensional constitutive equations for metal plasticity and numerical algorithms for solving nonlinear problems; the stress and strain analysis of a tensile bar can be extended beyond the point of necking; when the stress state becomes triaxial and deformation localises in the necking section (see curve "von Mises" in Fig. 8.8), and a "true" stress-strain curve of the material at higher deformations can be deduced.

Nevertheless, the deformation of the bar is not limited in the numerical model, whereas the tensile bar breaks after more or less plastic deformation in the real experiment (see last test point in Fig. 8.8). The reason is that the classical theory of plasticity by von Mises, Prandtl, and Reuss includes strain hardening of the material but no "strain softening" due to the evolution of damage in the material. Without going into details of the theory of ductile damage, the basic idea is that microvoids develop and grow under large plastic deformations, which cause a decreasing stress carrying capacity and finally result in the fracture of the bar (Gurson 1977; Tvergaard 1982). Introducing an additional variable characterising the damage of the material by the volume fraction of microvoids, modifying the yield condition for "porous" metals (Gurson 1977), establishing an evolution equation for the damage variable, and identifying all the additional material parameters of this Gurson, Tvergaard, and, Needleman model finally allows for a full description of the quasistatic mechanical behaviour of a tensile bar (Needleman et al. 1984; Tvergaard et al. 1984), see curve "Gurson" in Fig 8.8. The additional theoretical and numerical effort is considerable, but the profit is as important: The information which can be obtained in a simple tensile test has been extended significantly towards a unique characterisation of ductile material, including defor-

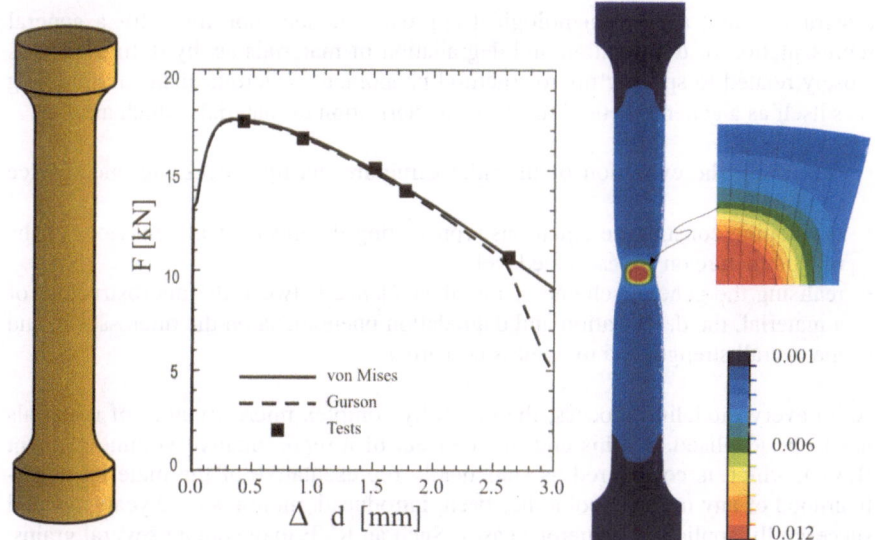

Fig. 8.8. Numerical simulation of a tensile test by the FE method; classical theory of plasticity by von Mises, Prandtl, and Reuss and porous metal plasticity by Gurson, Tvergaard, and Needleman; left: undeformed geometry of the specimen; centre: load, F, vs. reduction of diameter, Δd; right: deformation and damage distribution in the model

mation and damage. Because of that, the respective characteristic parameters can be transferred to simulate much more complicated loading configurations and phenomena like fracture of a cracked body or the Charpy impact test (Böhme et al. 1992; Brocks et al. 1995).

8.3.2 Micromechanical Modeling

As already mentioned above, continuum mechanics deals with idealised materials consisting of material elements, assuming that the material properties, the stresses and the strains within an infinitesimal neighbourhood of any material point can be regarded as uniform. On closer examination; however, matter is discretely structured and any material element consists of various constituents with differing properties and shapes, i.e., it has its own complex and, in general, evolving microstructure. Hence, the stress and strain fields within the material element likewise are not uniform at the microscale level. For describing the mechanical behaviour of materials with strong local gradients of the microstructure, the classical continuum approach is in general not sufficient.

Micromechanics has been established as a new and expanding field of research in materials science and constitutive modeling. One of its main objectives is to express the continuum quantities associated with a material element in terms of the parameters that characterise its microstructure and the properties of its microconstituents (Brulin et al. 1981; Nemat-Nasser et al. 1993). Micromechanical modeling takes place on a "meso level" between a full representation of the microstructure and a phenomenological approach. It does not allow for a general representation of deformation and degradation of materials as, by definition, it is closely related to specific microstructural phenomena. Micromechanical modeling sees itself as a general method for the characterisation of materials which aims at:

- describing the evolution of the microstructure during processing and service conditions,
- developing constitutive equations representing the mechanical behaviour of the microstructure on a mesoscale level,
- realising the general scheme of interdependence between the microstructure of a material, the deformation and degradation phenomena on the micro-scale, and the overall strength and toughness properties.

As in every modeling process, the generally complex microstructure of materials has to be idealised. To this end, the concept of a representative volume element (RVE), which is considered as statistically representative of the material neighbourhood of any material point, has been introduced more than 30 years ago and successfully applied in numerous cases. Such an RVE may contain several grains, different phases, inclusions, voids, cracks, and other micro-defects. In order to be representative, it must in general include a very large number of such micro-heterogeneities. However, the number of involved parameters will seriously interfere with a systematic treatment of the problem. It is therefore necessary to set up models as simple as possible but, nevertheless, exhibiting the characteristic microstructural phenomena.

A common simplification in micromechanical modeling is to assume a periodic microstructure; see Fig. 8.9, consisting, e.g., of hexagonal RVEs or "unit cells" containing an internal defect like a microvoid or microcrack. The hexagonal cells can be furthermore approximated by circular cylinders to allow simple axisymmetric FE calculations. Mesoscopic stresses and strains are defined by averaging the microscopic quantities over the cell volume and constitutive equations, for the material behaviour describing the average structural behaviour of the cell can be established on a "mesoscale." With a model like this, the influence of the inelastic matrix behaviour can be seperated from the effects of internal defects.

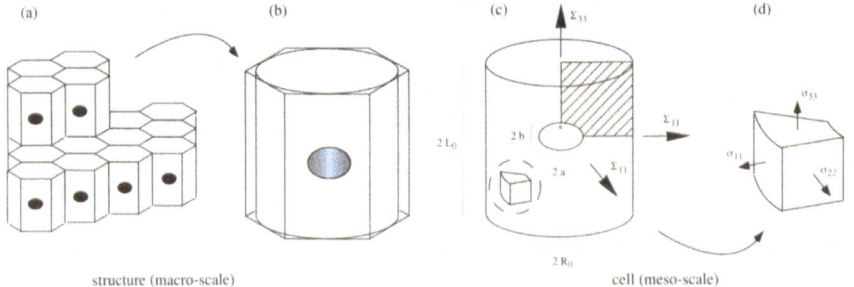

Fig. 8.9. Modeling the "continuum" (a) by a periodic assemblage of unit cells (b); visualising the different levels of scaling: macro ⇔ meso (a ↔ b), meso ⇔ micro (c ↔ d)

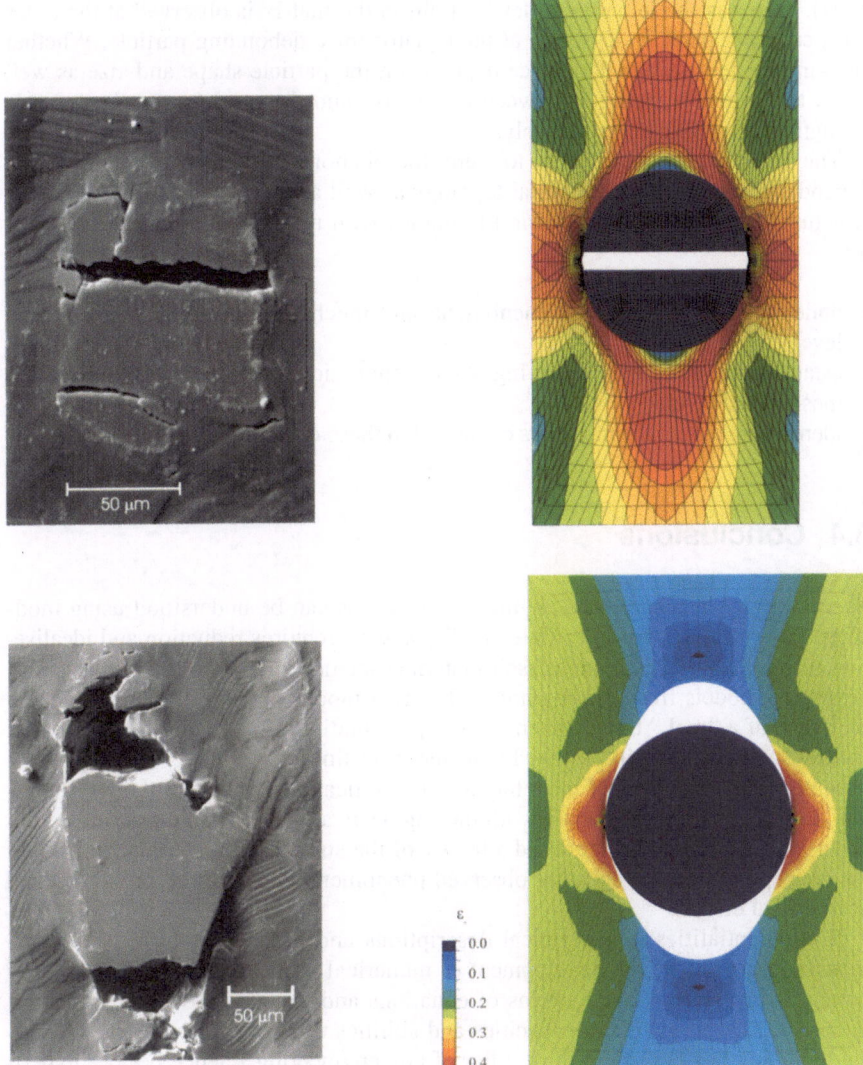

Fig. 8.10. In-situ observations (left) and FE modeling (right) of void nucleation at a brittle particle in a ductile aluminium matrix by particle cracking (upper) and particle debonding (lower)

Ductile tearing of metals, for example, is dominated by the mechanisms of void nucleation at particles, void growth and coalescence. Depending on the mechanical properties of the particle and the embedding matrix, void nucleation may occur either by particle debonding or particle cracking. In-situ observations in the scanning electron microscope allow to observe both mechanisms, see Fig. 8.10, left. These mechanisms can be modeled by unit cells containing particles. In the right column, Fig. 8.10 shows the respective FE representations of the two nucleation mechanisms and the fields of plastic strain around the particle (Steglich et al.

1997). Whereas the maximum plastic strain in the matrix is observed at the poles for a cracking particle, it occurs at the equator for a debonding particle. Whether cracking or debonding takes place depends on the particle shape and size as well as on the cohesive strength between the matrix and the particle and the fracture strength of the particle, respectively.

The numerical models allow to study the phenomena of damage evolution in dependence on the microstructural topology as well as on the interaction of microstructural constituents. Parametric FE studies with micromechanical models help in:

- understanding mechanical phenomena and mechanisms on a microstructural level,
- establishing, or at least justifying phenomenological constitutive equations on a mesoscopic level,
- identifying material parameters contained in these constitutive models.

8.4 Conclusions

Every activity in natural and engineering sciences can be understood using modeling, since dealing with a complex reality always requires reduction and idealisation of problems. This is the reason materials science and engineering mechanics have used models from the beginning. The first models, however, were models in the sense of a "real," three-dimensional representation of a structure, usually on a smaller scale. This is still practised in materials testing.

A new understanding of modeling as a theoretical description of the way a system or process works has arisen with the engineers and mathematicians of the 18th and 19th century, who established a theory of the strength of materials, which laid the conceptual basis of splitting observed phenomena into material properties and geometrical effects.

The potentialities of theoretical descriptions and predictions have been enormously improved by the development of numerical procedures for solving highly nonlinear problems, large systems of equations and the respective computer programmes, as well as by the capacities and abilities of current computers. Numerical modeling has become a third pillar of any engineering science besides experiments and theory.

Classical materials testing, i.e., the determination of "characteristic parameters" on test specimens such as yield strength, ultimate tensile strength, fatigue resistance, fracture toughness, etc., faces many problems of transferability of these parameters from the specimen under commonly uniaxial loading and uniform stress and strain fields to a real component under multiaxial loading and stress and strain gradients.

A modern concept of materials characterisation has to be based on all three "pillars" of modeling, i.e. testing, constitutive theories and numerical simulations. A material characterisation is meaniful only in the context of "constitutive models," which describe the mechanical behaviour. This requires an interdisciplinary cooperation of experts in materials science, continuum mechanics, and numerical analysis.

The eternal as well as silly question which has to be given priority, experiment or theory, has already been answered by Immanuel Kant (1724 – 1804) more than 200 years ago: "Anschauungen ohne Begriffe sind blind, Begriffe ohne Anschauungen sind leer" (perceptions without conceptions are blind, conceptions without perceptions are empty).

Chapter 9
Mathematical Morphology

by Michel Schmitt

Abstract

The structural analysis of an image and the recognition of certain objects in it generally require two steps: one consists in identifying the structures of interest within the image and isolating them (this we call segmentation), and the second one in quantifying thes objects by associating values to them (numbers or symbols) with a view to their classification.

In order to explain these two steps, we begin by examining the tools at our disposal and their manual or automatic organisation for segmentation tasks. We then show how these same tools enable us to mesure the segmented objects.

9.1 Introduction

Mathematical morphology is a mathematical theory based on set representation, geometry and algebraic structures, used mainly in image processing. The first steps were initiated in the 70's by Matheron (1967, 1975) and Serra (1982, 1988) for quantifying sizes in porous media and more generally shapes, usually modeled by random sets. In the 80's, the theory extended to functions (Serra 1982, Sternberg 1986) enabling the study of gray scale images. The applications ranged from biology, material science and remote sensing to automatic control and many others. More recently, the general algebraic framework of lattices has been set (Matheron 1983, Serra 1988, Heijmans 1994), allowing the theory of morphological filtering to emerge. In parallel, the watershed transformation has been proposed (Maisonneuve 1982), along with a very efficient algorithm (Vincent 1991), which has been applied worldwide to image segmentation tasks. Today, these algorithms are the basis of the new generation of image coding norms (Mpeg 4) (Salembier et al. 1995).

This paper is an illustration of the tools provided by mathematical morphology, intended for image processing practitioneers. Many theoretical aspects have been left aside, in order to enhance the tools and their effects on images. We present first the four classical operators (dilation, erosion, opening and closing) and illustrate their usage on image filtering. Then the watershed is described, along with its segmentation properties. Finally, we investigate the original way images can be quantified, in the deterministic as well as in the random case.

9.2 An Introductive Example

Figure 9.1a is an image of coffee beans. The question is: how many beans are present on this image? Counting the number of black objects is not sufficient, because some beans are overlapping. If we consider the set of black pixels, a characteristic feature of the overlap is that the set becomes narrower between the objects. In order to detect these narrowings, suppose we have at our disposal a disk of radius r which fits inside one bean. If the diameter of the disc is larger than the neck between two beans, then the disk cannot be translated from one bean to a neighbouring bean without going outside the set of beans. In Fig. 9.1b, all the positions where a disk fits inside the beans are depicted. We see that the beans have shrunk (a strip of width r has been removed around each bean), some have disappeared (the disk is too large to fit inside), and some have been disconnected (this is what we want). This set transformation has been called *erosion by a disk*.

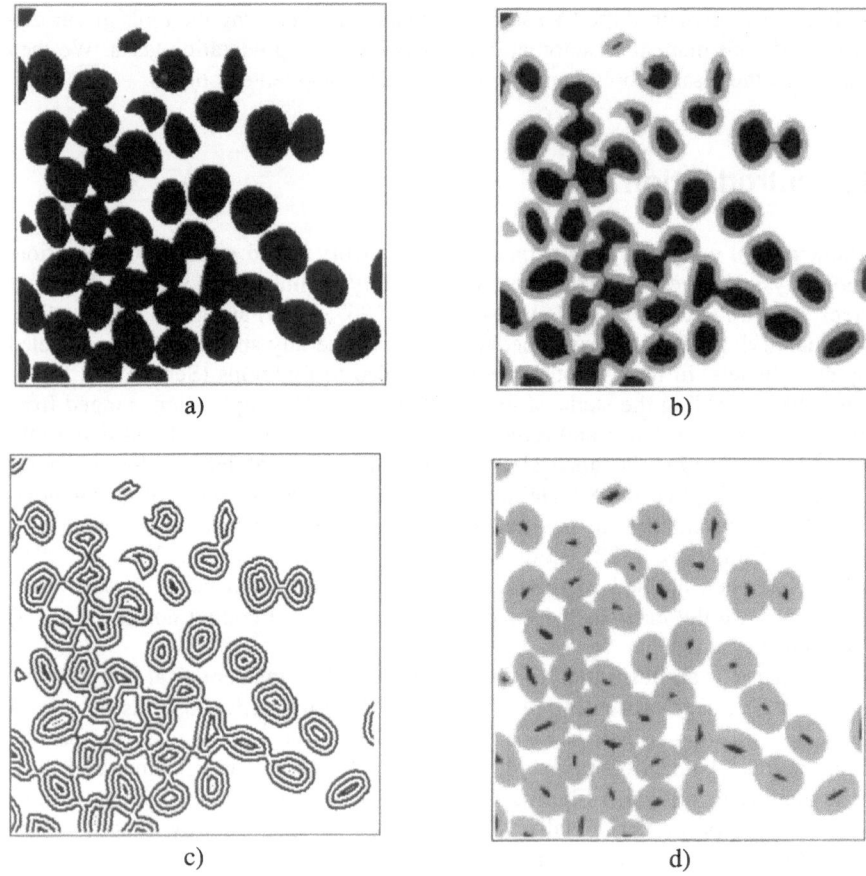

a)   b)

c)   d)

Fig. 9.1. (a) How much coffee beans? (b) Separating some of the beans by erosion. (c) Distance function and separation at any size. (d) Centers of the beans

However, a unique radius of the disk is not sufficient to do the job. So, all the sizes of disks are used, and as the result is decreasing according to inclusion as the radius becomes larger, we can pile up all the sets, giving rise to the function depicted in Fig. 9.1c. Finally, the centers of the beans are the local maxima of this function (Fig. 9.1d).

This toy example illustrates many facets of mathematical morphology:

- The problem has been solved using geometrical (disk) and set concepts (inclusion).
- The unknown image has been investigated by a completely known object (the structuring element); here a disk.
- The useful features of the problem (narrowing, convexity of the grains) have been translated into image transformations (erosion, local maxima), extracting these features.
- A unique transformation is not sufficient: many erosions have been used in conjunction with local maxima: a morphological program is usually a sequence of transformations.

9.3 The Morphological Tool Box

The most illustrative framework for the operators provided by mathematical morphology is geometry on sets. So, the images we are dealing with have to be sets. We usually handle the following two types of images:

- A *binary image* is a function from a rectangular domain that can only take two values, which, by convention, are equal to 0 or 1. When we depict binary images, the pixels with a value 1 are printed in black, while those with a value 0 are white. The associated set is composed of the pixels with value 1.
- A *gray scale image* or *digital image* is a function f from a rectangular domain, with real or discrete values as the case may be. It is described with its level sets $X_\lambda = \{x \mid f(x) \geq \lambda\}$. The sequence X_λ is decreasing with λ.

So, an image transformation working on binary images may be extended to gray scale images simply by "piling up" all the transformed level sets. In order to do so, the binary operator has to be increasing, *i.e.*, it must transform the decreasing sequence of sets into another desceasing sequence (see Fig. 9.2).

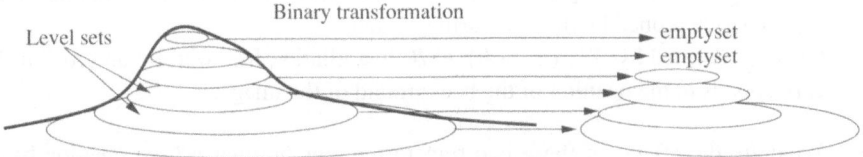

Fig. 9.2. Representation of a function by its level sets. Each level set is transformed and its piling-up gives the transformed function

9.3.1 The Four Operations

Mathematical morphology is based on the solid principle of comparing the unknown structure, *i.e.*, the image under study, with a set of shapes, the structuring element, which is completely known. Boolean relationships such as intersection or inclusion are used as the comparison method.

In theory, all the morphological operators can be derived from a single operator or a primitive operator, namely dilation.

The dilation of the image X by the structuring element B, denoted $X \oplus B$, is the set of the positions u, for which B_u, translated into u of B, intersects X:

$$X \oplus \breve{B} = \left\{ u,\, B_u \cap X \neq 0 \right\} = \left\{ x - b,\, x \in X \text{ and } b \in B \right\} \tag{9.1}$$

In accordance with the morphological principle, the Boolean relationship employed here is the intersection. The extension to function by the piling-up principle yields an image function f:

$$f \oplus \breve{B}(x) = \sup\left\{ f(y),\, y \in B_x \right\} \tag{9.2}$$

Then, by duality, we can define the **erosion** as the dilation of the background of the image:

$$X \ominus \breve{B} = (X^c \oplus \breve{B})^c = \left\{ u,\, B_u \subseteq X \right\}$$

and   (9.3)

$$f \ominus B(x) = \inf\left\{ f(y),\, y \in B_x \right\}$$

Erosion corresponds to the inclusion of the structuring element. An application of the erosion has been presented in the preliminary example. This duality principle is at the basis of the algebraic theory of morphology: two operators in adjunction form an erosion-dilation pair.

The fundamental feature of the already defined two operators is that they are not inverse from each other. Iterating one with the other gives rise to new operators, which have different properties:

- **Opening**, defined by X_B $(X \ominus \breve{B}) \oplus B$, is an erosion followed by a dilation. It fits the space probed by the structuring element.
- **Closing**, defined by X^B $(X \oplus \breve{B}) \ominus B$, is a dilation followed by an erosion. It corresponds to the opening of the background of the image.

What are the effects of these two transformations on images? An opening by a disk eliminates:

- small objects which do not contain the structuring element (the disk) in their interiors,
- narrow parts of objects,
- small protrusions on the boundary.

A closing by a disk:
- fills small crevices on the boundaries of objects,
- fills holes which do not contain the structuring element,
- clusters small objects (eliminates narrow parts of the background).

On gray scale images, the effects are similar; changing "objects" by bright zones, "background" by dark zones.

For example, openings may be used to discriminate objects according to their size (see Fig. 9.3). On gray scale images, the opening removes bright lines or objects, so that subtracting the opening from the original image allows to extract these features. This transformation, called **Top-Hat** is illustrated in Fig. 9.3.

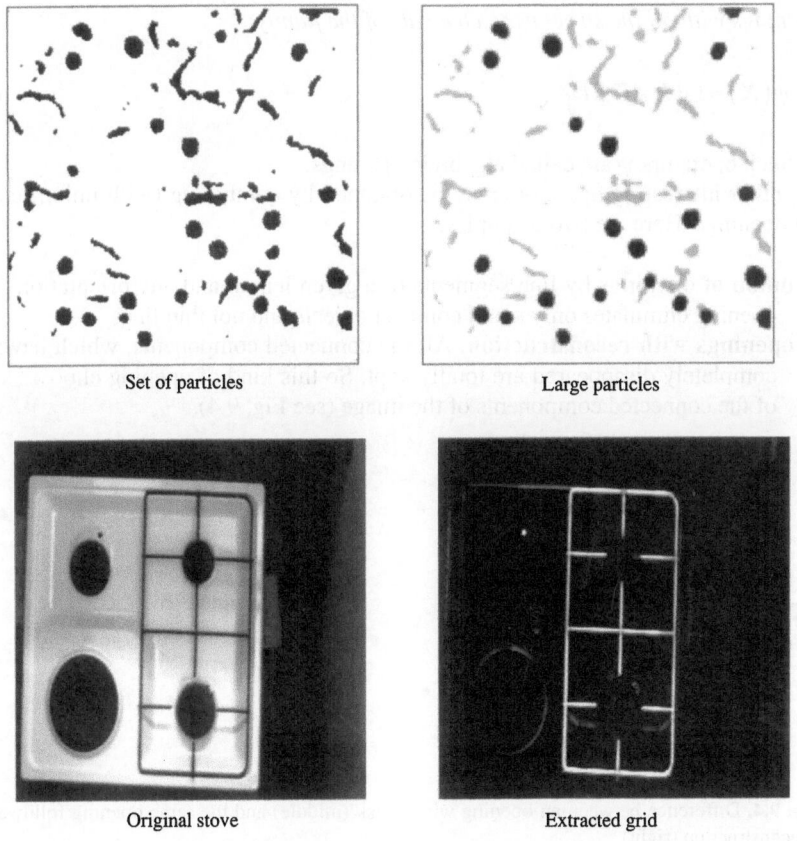

Set of particles   Large particles

Original stove   Extracted grid

Fig. 9.3. Examples of applications with openings by a disk. Top row: selection of objects according to size. Bottom row: extraction of linear bright features by top-hat

9.3.2   Characterisation of Openings and Filtering

This construction of operators seems a little arbitrary. However, all the openings may be characterised algebraically, showing that the construction we have presented is complete. To this purpose, we need some mathematics.

An opening ψ has four main mathematical properties:

- **Increasingness:** it preserves the inclusion, i.e., $X \subseteq Y \Rightarrow \psi(X) \subseteq \psi(Y)$.
- **Idempotence:** nothing is gained by iteration, i.e., $\psi(\psi(X)) = \psi(X)$.
- **Anti-extensivity:** the result is smaller than the original image, i.e. $\psi(X) \subseteq X$.
- **Translation invariance:** it commutes with translations, i.e. if τ is a translation, $\psi(\tau(X)) = \tau(\psi(X))$.

The following theorem shows that these four properties nearly characterise the openings (Matheron 1986).

Theorem 2.1. *If an operator ψ is increasing, idempotent, anti-extensive and commutes with translation, then there exists a family of structuring elements B (and not one structuring element usually) such that ψ is the union of all the openings with all the structuring elements of the family:*

$$\psi(X) = \cup\left\{X_B \middle| B \in B\right\} \tag{9.4}$$

Such operators ψ are called algebraic openings.

So, other interesting operators may be obtained by combining (with union) classical openings. Here are two examples:

1. **union of openings** by line segments of a given length and any orientation. This opening eliminates only small compact objects and not thin lines.
2. **openings with reconstruction.** All the connected components, which have not completely disappeared are totally kept. So this kind of opening chooses some of the connected components of the image (see Fig. 9.4).

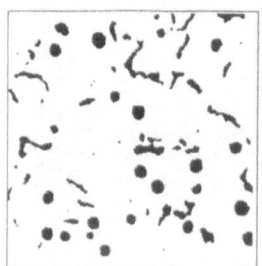 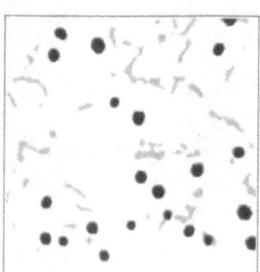 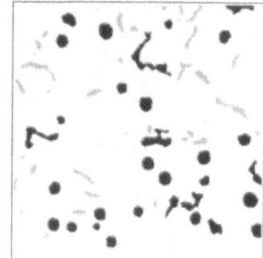

Fig. 9.4. Difference between an opening with a disk (middle) and the same opening followed by a reconstruction (right)

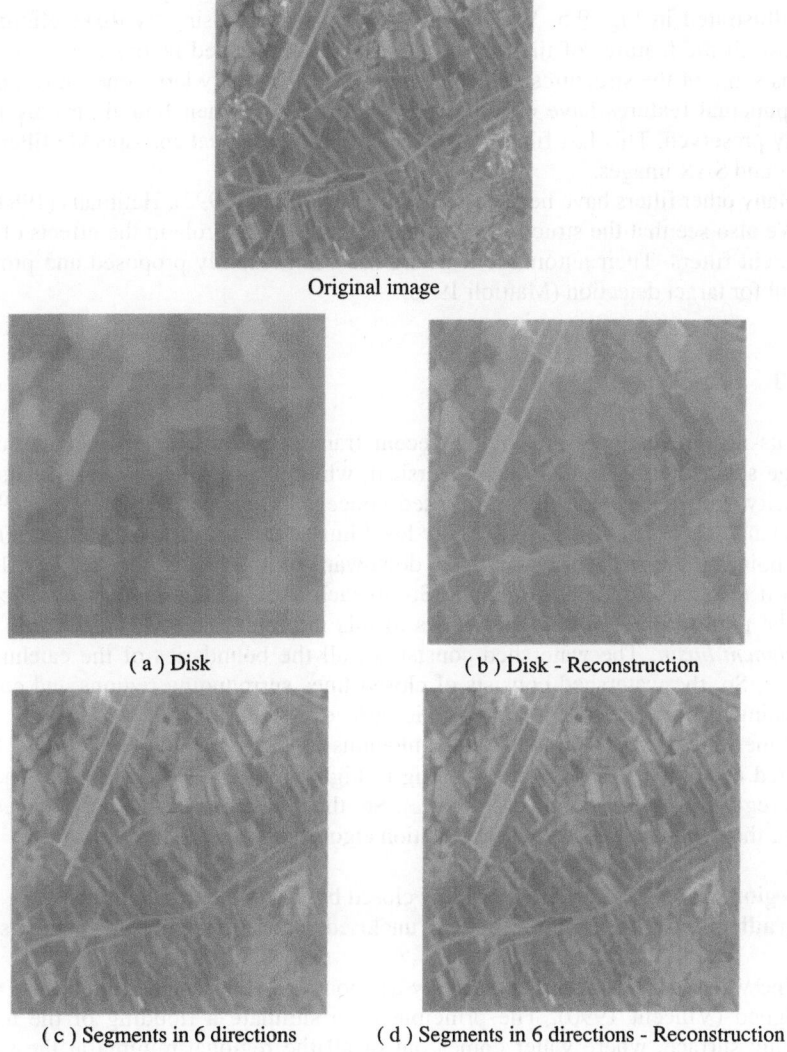

Original image

(a) Disk   (b) Disk - Reconstruction

(c) Segments in 6 directions   (d) Segments in 6 directions - Reconstruction

Fig. 9.5. Different alternating sequential filters

The same kind of statement exists for closings and algebraic closings. If we iterate openings and closings with larger and larger structuring elements, we obtain a a new class of operators called **alternating sequential filters** (FAS for short). They are no more anti-extensive, but remain increasing, idempotent and commute with translations. These filters are commonly used to filter out some noise, especially speckle on radar images. To understand the variety of filters, let us enumerate some openings we can use: openings by disks, openings by disks with reconstruction, union of openings by line segments and union of openings by line segments with reconstruction. Then the range of sizes has to be adjusted. These filters are illustrated in Fig. 9.5. We see that openings and closing by disks eliminate almost all the features of the original image (a). The added reconstruction in (b) keeps some of the structures, which are connected to a very large one. In (c), only the punctual features have disappeared, whereas in (d) their boundaries are perfectly preserved. This last filter is the basis of morphological anti-speckle filters in radar and SAR images.

Many other filters have been described in Sierra et al. (1992), Heijmans (1984).

We also see that the structuring elements play a crucial role in the effects of the different filters. Their automatic learning has been recently proposed and proven useful for target detection (Mattioli 1996).

9.3.3 Watersheds

Let us now investigate one of the recent transformations, extensively used in image segmentation, namely the watershed, which is so called because of its similarity with the geographic watershed concept (Vincent 1990, Beucher 1990, Najman et al. 1993). Consider the gray level image as a topographic surface. From any point on the surface, water flows downward on the line of the steepest slope until it reaches a regional minimum. So, to each regional minimum, we associate all the points from which water flows to this minimum and call this region the *catchment basin*. The watershed consists of all the boundaries of the catchment basins. So, the watershed consists of closed lines surrounding regions and corresponding to crest lines of the topographic surface.

If the watershed is computed on the modulus of the gradient of an image, it is located on the crest lines corresponding to high contrast pixels (contours) enclosing regions of lower contrasts (objects). So, the watershed has the advantages of the of the two main classes of segmentation algorithms:

1. **region growing** algorithms, yielding closed but badly located contours,
2. **gradient based** algorithms, yielding unclosed but accurately located contours.

The watershed was used extensively as soon as a very efficient algorithm was designed (Vincent 1990). The principle is to simulate a flooding of the topographic surface, where water comes out of all the regional minima at the same altitude. The lines where waters coming from different minima meet build the watershed. This kind of implementation eliminates the difficulties associated with the digital nature of images and especially the problem of flat zones.

Used on the inverted distance function of the coffee beans in Fig. 9.1c, the watershed precisely gives the separations between the grains. Note that the centers of the grains are precisely the regional minima of the inverted distance function, and each center gives rise to a catchment basin corresponding to the grains (see Fig. 9.6).

If directly used on the modulus of the gradient on a real image, the watershed gives disappointing results: due to the natural level of noise on images, the watershed exhibits an oversegmentation (see Fig. 9.7a and b). However, the strength

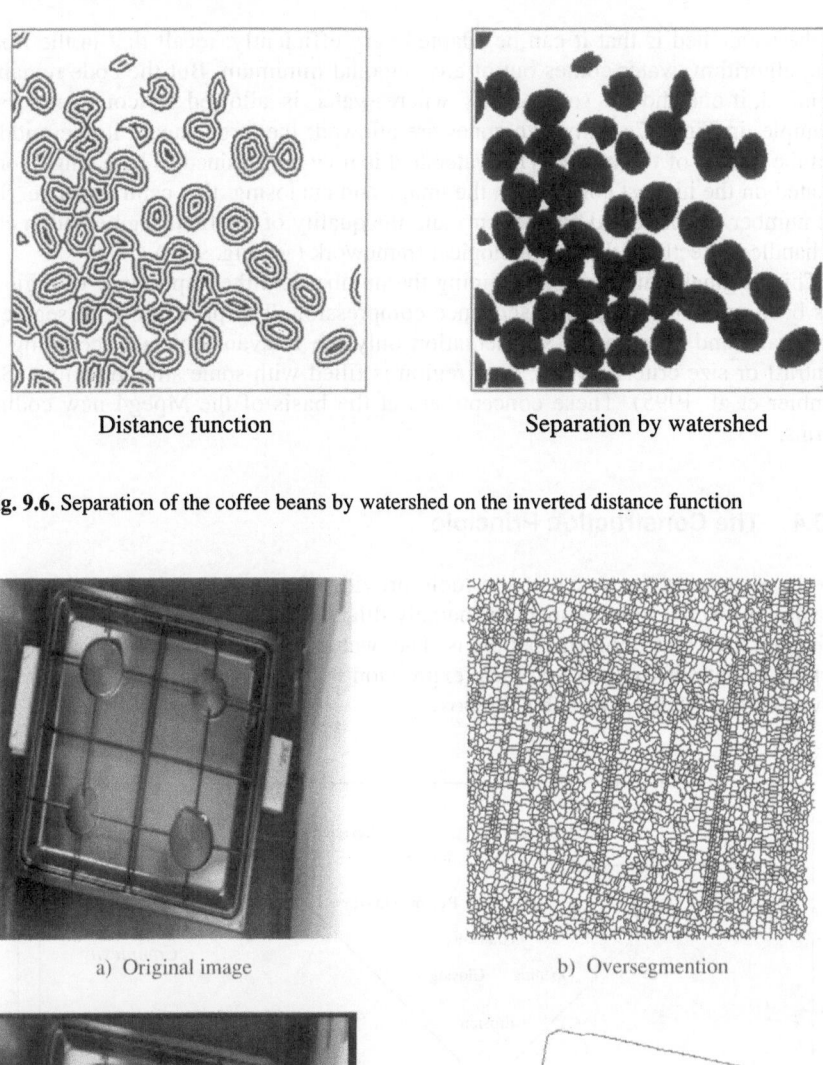

Distance function   Separation by watershed

Fig. 9.6. Separation of the coffee beans by watershed on the inverted distance function

a) Original image   b) Oversegmention

c) Imposed minima   d) Constraint watershed

Fig. 9.7. Watershed segmentation: the problem lies in extracting the outer boundary of the cookstove grid

of the watershed is that it can be adapted very efficiently: recall that in the flooding algorithm, water comes out of any regional minimum. But the code remains
identical, if one chooses some zones where water is allowed to come out. For
example, in Fig. 9.7c, only two zones are allowed: the large square in the middle
and the border of the image. The watershed is now constrained to be a single loop
located on the highest contrasts in the image and enclosing the central square. So,
the number of connected components and the quality of the final segmentation can
be handled directly in the morphological framework (see Fig. 9.7d).

This original feature of constraining the number and the importance of regions
has been used for image and sequence compression. The principle is to segment
the image and keep in the segmentation only the relevant objects, according to
contrast or size criteria. Then, each region is filled with some simple texture (Salembier et al. 1995). These concepts are at the basis of the Mpeg4 new coding
norms.

9.3.4 The Construction Principle

We have seen so far some of the tools provided by mathematical morphology.
Stemming from a unique operator, namely dilation, all the others can be derived,
using iterators and Boolean operations. The watershed itself can be expressed in
terms of dilations, but the complete expression is too long to be explicitly written
down. Figure 9.8 illustrates this toolbox.

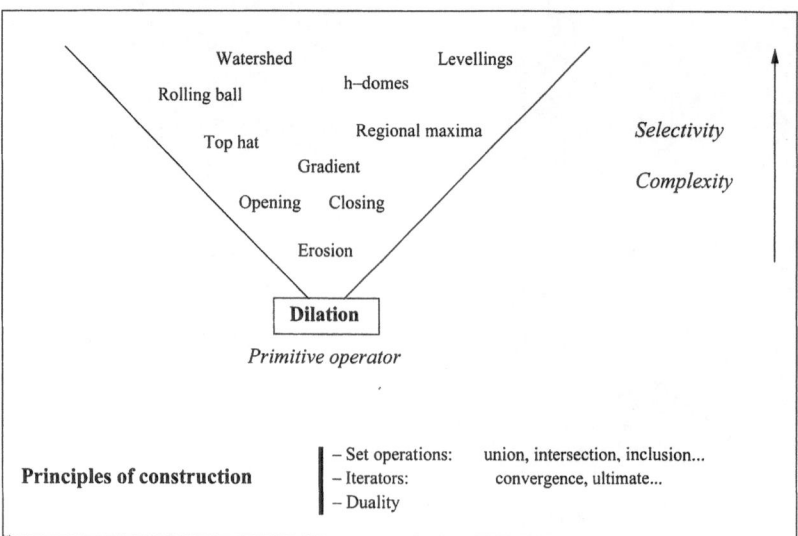

Fig. 9.8. The structure of morphological transformations and their construction principles

9.3.5 Segmentation Programme

Now, how do we combine the morphological operators in order to segment one or several objects? The choice of the transformations is influenced by the necessary and sufficient characteristics which enable us to determine the interesting structures from the other ones. In practice we will have to do with sufficient characteristics only. It is worth noting here that these characteristics are not only associated with the object we want to segment, but also with the image background from which we want to distinguish it. This is known as the "universe" by researchers in artificial intelligence.

There is another reason why the characteristics can only be sufficient: an image analysis problem has numerous solutions as can be seen in Schmitt (1991) giving more than ten different solutions to the problem of lamellary eutectics (Fig. 9.9). Each one of the solutions is **minimal** in the sense that in a given universe, eliminating one characteristic involved in the solution will give incorrect results in certain cases.

The choice of one or the other solution is thus made according to two contradictory criteria. The first one maintains that the best solution is the one that functions in the largest possible universe, whereas the second one measures the efficacy of the solution in terms of running speed, knowing that certain realistic hypotheses can be made on the universe: a lamellary eutectics image will never enclose an elephant.

In contrast to an approach of the neural network type that requires an important image base: (hand-written character recognition is only solved with a database of several thousand characters (Le Cun 1987)) the mathematical morphology approach only requires a limited number of representative images in order to solve the problem. On the other hand, the prerequisite is a minute analysis of the problem and a clear definition of the characteristics involved in the solution. In this way, the cause of any failure is much easier to detect. The morphological program

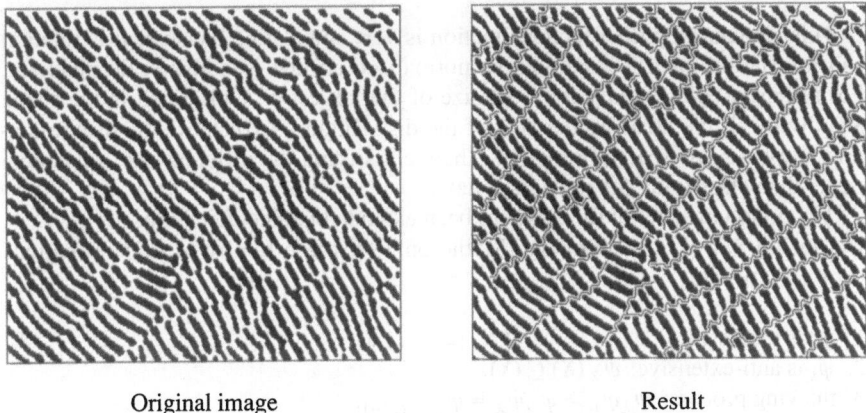

Original image   Result

Fig. 9.9. Problem of lamellary eutectics: find the default lines

is no longer a black box: it clearly shows the necessary characteristics of the image in the resolution process of the problem.

9.4 Quantification and Morphological Measurements

After the segmentation step, the quantification step makes it possible to measure the objects of interest with a view of classifying them. Thus, quantification reduces and structures the information considerably, which considerably simplifies the classification. For this task, mathematical morphology has numerous tools at hand.

For an almost exhaustive study of morphological measures, see Serra (1982) and Coster et al. (1985). We present here only two examples: granulometries and random models.

9.4.1 Granulometries, Spectral Function and Curve by Erosion

Here, our goal is to quantify shapes on binary images. If $B(r)$ denotes the disk of radius r, the opening with $B(r)$ eliminates all the parts which are narrower than $2r$. So, the difference between the two openings by $B(r)$ and $B(r+1)$ is precisely the part whose width is $2r$ or $2r+1$. If we plot the area of the opening $X_{B(r)}$ versus r, we get a kind of signature of our shape. If we also want to quantify the sizes of the concavities of the shape, we can plot the area of the closing $X^{B(r)}$ versus r.

Definition 3.1: *Let X be a shape. The* spectral function $\sum_X$ *of X is defined by:*

$$\sum_x(r) = \begin{cases} A(X_{B_{(r)}})/A(co(X)) & \text{if } r \geq 0 \\ A(X^{B^{(r)}})/A(co(X)) & \text{if } r < 0 \end{cases} \tag{9.5}$$

where $co(X)$ is the convex hull of X, $A(X)$ the surface of X.

This signature by the spectral function is very interesting, because it is rotation and translation invariant and can be normalised so that it is also scale invariant. For instance, take as unity for r the size of the largest disk inscribed in the shape and divide all the areas by the area of the disk. Figure 9.10 shows the discriminating power of this spectral function. Then, classifying the different curves is much easier than classifying the raw image data.

The notion of spectral function has been extended in algebraic terms as:

Definition 3.2: A granulometry (Matheron 1975) is a family of transformations $(\psi_\lambda)_{\lambda \geq 0}$ verifying:

1. ψ_λ is increasing: $X \subseteq Y \Rightarrow \psi_\lambda(X) \subseteq \psi_\lambda(Y)$,
2. ψ_λ is anti-extensive: $\psi_\lambda(X) \subseteq (X)$,
3. thieving process: $\psi_\lambda \psi_\mu = \psi_\mu \psi_\lambda = \psi_{\max(\lambda,\mu)}$.

The granulometric curve associated with granulometry

$$\lambda \to \Psi_X(\lambda) = A\big(\psi_\lambda(X)\big) \tag{9.6}$$

The associated inverse problem, *i.e.*, represent the shapes having the same spectral curve, has not yet been solved. However, what has been solved is the curve by erosion case (surface of the eroded images of a set in terms of the radius of the structuring element disk (Mattioli et al. 1992). Clearly, two identical shapes up to a translation, rotation and scale factor have the same curve by erosion, but the converse is not true. Let us take this a little further: a shape is represented by its skeleton (centre of the maximal balls included) and its quench function (radius of the maximal ball). This quench function has four types of interesting points (Fig. 9.11).

- local maxima on an arc in the skeleton (protrusion),
- local minima on an arc in the skeleton (narrowing),
- local maxima on a triple point in the skeleton,
- the triple points in the skeleton, without there being local maxima.

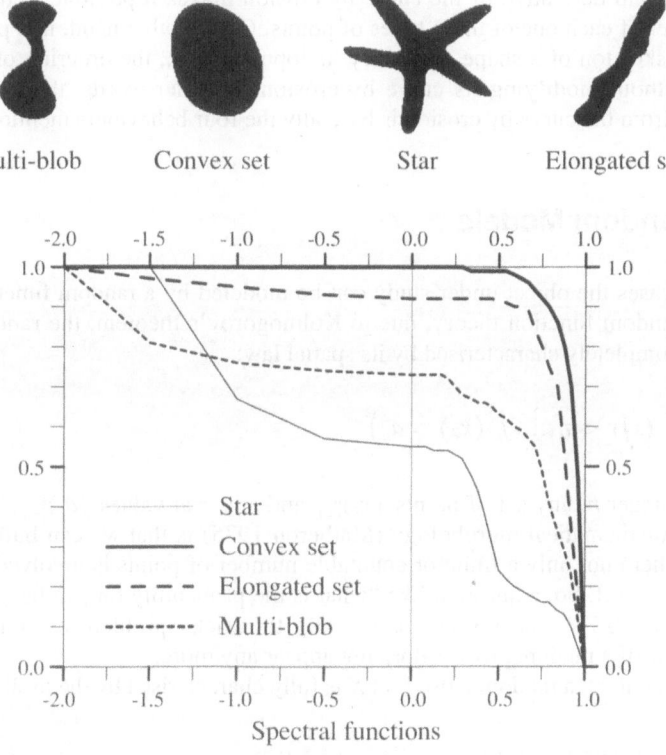

Fig. 9.10. Example of shapes and their spectral function

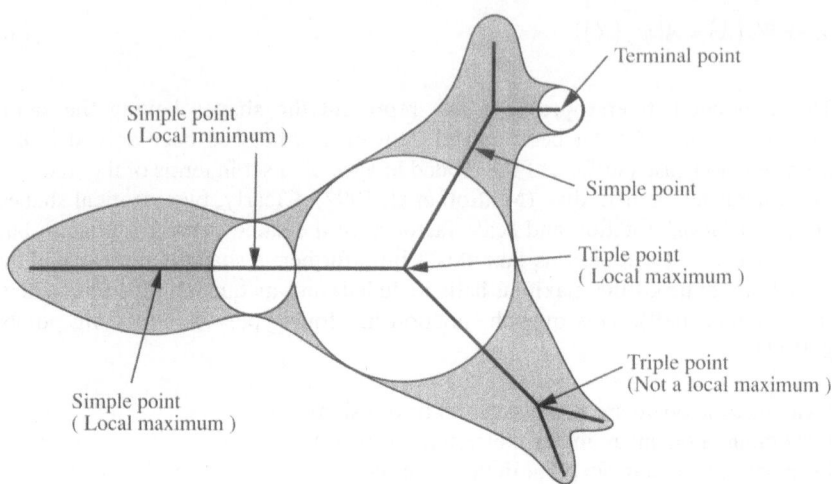

Fig. 9.11. Skeleton of a shape and characteristic points in the skeleton

The second derivative of the curve by erosion makes it possible to distinguish the number of each one of these types of points. On the other hand, it is possible to twist the skeleton of a shape or modify its topology, i.e., the ordering of its triple points without modifying its curve by erosion. In other words, the information obtained from the curve by erosion is basically the four behaviours mentioned.

9.5 Random Models

In some cases the object under study can be modeled by a random function. Usually, in random function theory, due to Kolmogorov's theorem, the random function f is completely characterised by its spatial law:

$$Pr\left(f\left(x_1\right) < a_1, \ldots, f\left(x_n\right) > a_n\right) \tag{9.7}$$

for any integer n, any set of points $(x_i)_{i=1}^n$ and any real values $(a_i)_{i=1}^n$. The originality in mathematical morphology (Matheron 1975) is that we can build another theory, where not only a finite or countable number of points is involved, but also any compact set. So, a question like "what is the probability for f to be positive on the interval [0,1]" makes sense. Especially the tricky problem of continuity of realisations of a random process does not appear any more.

In this context, a random closed set X is fully characterised by the probabilities

$$T\left(K\right) = Pr\left(X \cap K \neq 0\right) \text{ for any compact set } K \tag{9.8}$$

(Choquet's theorem (Matheron 1975)).

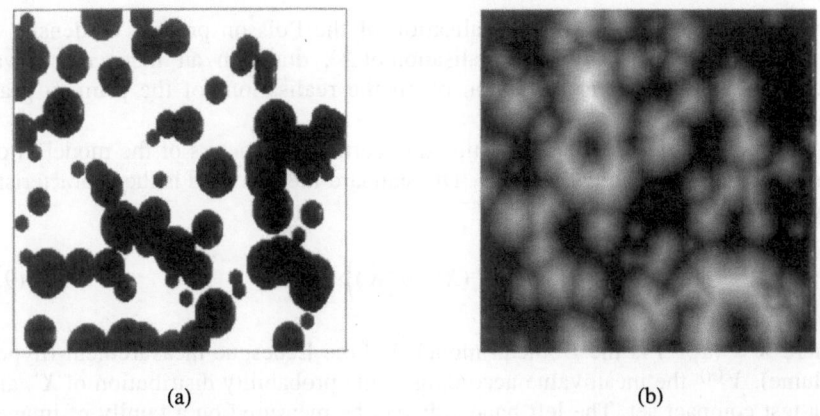

(a)   (b)

Fig. 9.12. (a) Boolean model $X \equiv (a, X') \equiv X = \cup \{X'_x, x \in P\}$, where P is a Poisson process of density a and X' is a disc of random radius. (b) Extension of the model to digital functions

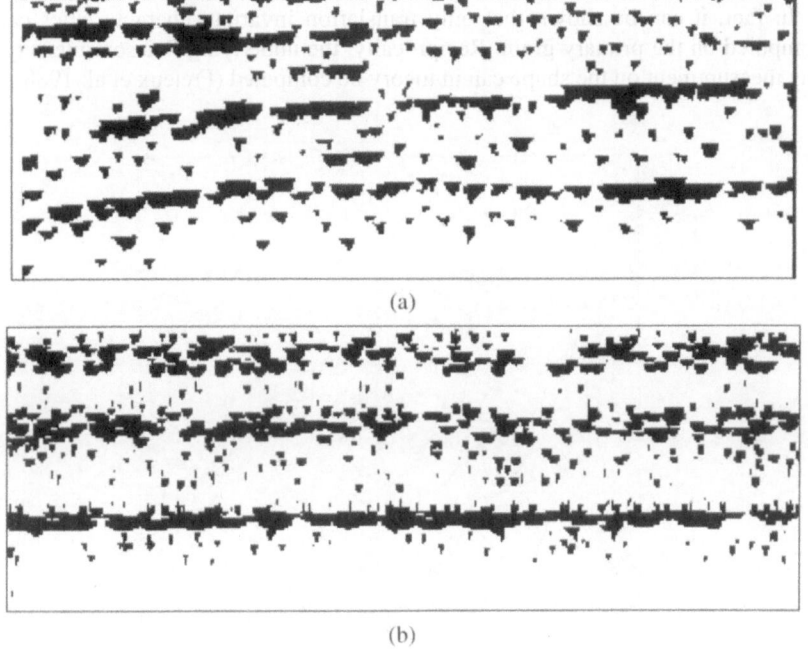

(a)

(b)

Fig. 9.13. Oil reservoir modeling by a Boolean model. (a) True data corresponding to an outcrop. (b) Simulated data with the inferred parameters

Let us now examine one particular model, namely the *Boolean model*. It randomly models the distribution of random objects drawn in an independent way from a base shape that we call *primary grain*. This model depends on a positive real number a, the density of the objects, and a random set X'. A realisation of this model is obtained by taking a realisation of the Poisson process of density a, translating at each of its points a realisation of X', drawn in an independent way each time and by taking the union of all the realisations of the primary grain (Fig. 9.12).

The problem thus consists in estimating certain parameters of the model in order to distinguish different textures. The standard method used is the characteristic equation of this model:

$$P\ (K\ \subset\ X^c)\ =\ \exp\left(-a\overline{V^{(n)}}\ (X'\ \oplus\ \check{K})\right) \tag{9.9}$$

where $X \equiv (a,X')$ is the Boolean model, $V^{(n)}$ the Lebesgue measurement (hypervolume), $\overline{V^{(n)}}$ the mean value according to the probability distribution of X', and K a test compact set. The left hand side can be measured on a family of images. Indeed, on account of the ergodicity of the model, only one image is necessary. The right hand side can be calculated in the case where $K=B\ (\lambda)$ is a ball of radius λ and X' a convex primary grain. In this way we can estimate a, the mean surface and the perimeter of X'.

In fact, it can be shown that only translation invariant characteristics can be computed on the primary grain. Reciprocally, the underlying Poisson intensity and any measurement on the shape can in theory be computed (Frèteux et al. 1988).

Fig. 9.14. Dead leaves model of stars. In contrast to the Boolean model, the objects fall sequentially and occlude all the underlying objects

The next step involves simulating the model with the inferred parameter, constrained by the available data, like punctual values, mean values over a field, connectivity constrains, or others (Lantuéjoul 1996, 1999).

Boolean models have been used in many applications, like oil reservoir modeling (Fig. 9.13 (Schmitt et al. 1996)).

Many other examples of random models exist, like Cox models and Dead leaves models (Fig. 9.14). For an extensive review, see Matheron (1975), Serra (1982), Stoyan et al. (1987).

Chapter 10
Statistical Interpolation Models

by Hans Wackernagel and Michel Schmitt

Abstract

Statistical interpolation methods for irregularly spaced data are reviewed in this paper, with a special focus on the geostatistical model.

Geostatistics subdivides the interpolation of spatial data into two steps: first an analysis and modeling of the average spatial variation, then a spatial regression on the interpolation node from sample locations nearby. With a principal variable and auxiliary variables, there are several ways to apply geostatistics:

1. ignoring the auxiliary information (kriging),
2. including it as a random quantity (cokriging), or
3. including it as a deterministic quantity (external drift).

These three strategies are compared with respect to different sampling configurations and coregionalisation models.

Splines have been often compared with kriging techniques. We give a simplified exposition of intrinsic random functions theory and of kriging with a translation-invariant drift. Kriging with a filtering of white noise is then shown to be equivalent to smoothing thin-plate splines.

10.1 Introduction

Geostatistics was initially coined (in the 1950s) as being an application of statistical methods to geology. Then, with the extension of its application fields to other branches of the earth sciences like oceanography or meteorology, it has been defined as the application of the theory of random functions to natural science. Nowadays, with geostatistics spreading into fields like material science or environmental analysis, the frame again needed to be enlarged, and geostatistics can be conceived as the application of the theory of random functions (or sets) to phenomena with a spatial support.

Geostatistics offers interesting extensions of *objective analysis techniques*, as they are called in meteorology (Gandin 1963; Chauvet et al. 1976; Thiébaux and Pedder 1987). Geostatistics can be used in combination with data assimilation (Daley 1991) and especially with Kalman filtering (Huang and Cressie 1996). Thiébaux (1997) has attempted a sketch of a unified framework for these different techniques. Other new developments of interest in climatological geostatistics are

Lagrangian kriging (Amani and Lebel 1997) and kriging in the space of empirical orthogonal functions for downscaling (Biau et al. 1999).

As stated by Kitanidis (1997), "the scope of geostatistics has gradually expanded to include applications that traditionally have been addressed using spline interpolation methods." Mardia et al. (1996) even qualify spline fitting as "a special case of kriging." Naturally, this process did not go its way without a reaction from the community of spline fans. At some stages, it has led to open polemics: Wahba (1990a) versus Cressie (1990); Hutchinson and Gessler (1994) versus Dubrule (1984). A summary is given in Cressie (1991, pp 180 – 183).

In this paper, we first present basic geostatistics and apply it to map Cadmium in a region of the Swiss Jura. We then examine kriging with drift as well as dual kriging in the framework of IRF-k theory (Matheron 1973), and then discuss the equivalence with smoothing thin-plate splines. The presentation draws largely on Wackernagel (1998), where a more detailed exposition is found.

10.2 The Random Function Model

A limited domain D of space (3D, 2D, 1D) or time is studied, constituted of points x. At each point $x \in D$, we construct a random variable $Z(x)$. The family of random variables $\{Z(x); X \in D\}$ is termed a random function. A realisation (draw) of the random function is called a regionalised variable $z(x)$.

In applications, we often only possess data (samples) about a single realisation of the random function. The epistemological problem of the inference of parameters of the random function from a single realisation is resolved by different hypotheses of stationarity and ergodicity, which are discussed in Matheron (1989).

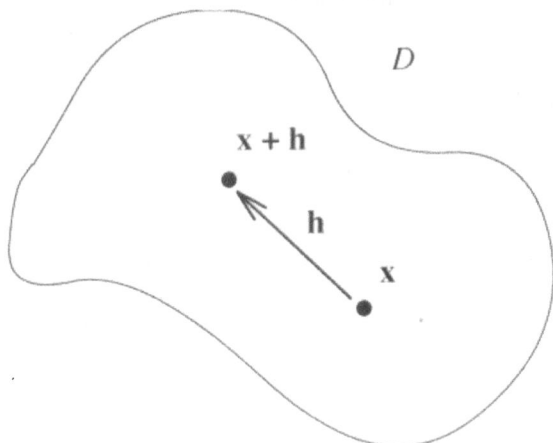

Fig. 10.1. A spatial domain D and two points of D connected by a vector **h**

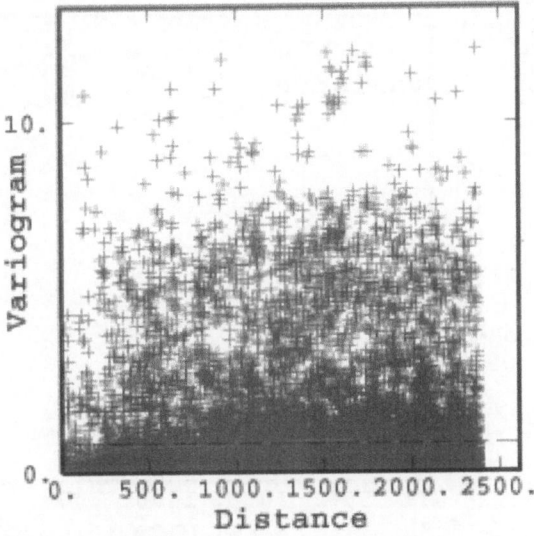

Fig. 10.2. Diagram of squared differences of pairs of data points as a function of distance in geographical space. The data are 259 Cadmium values (mg kg⁻¹) and the distance is in metres. The dashed line at the bottom of the graph represents the statistical variance

10.2.1 The Variogram

Let **h** be a vector separating two points in a domain D, as shown on Fig. 10.1.

We can compare the measured quantity of interest at two locations **h** apart by computing the dissimilarity measure:

$$\frac{\left(z(x+\mathbf{h})-z(x)\right)^2}{2} \tag{10.1}$$

where 1/2 is simply a normative constant.

These squared differences can be plotted against the distance |**h**| separating each pair of points. Such a *variogram cloud* is depicted on Fig. 10.2.

In the next stage of the analysis of spatial variation, averages are computed for distance (and angle) classes $\mathbf{h}_k$, as represented on Fig. 10.3. This is the *experimental variogram* $\gamma * (\mathbf{h}_k)$.

For linking the experimental variogram to a random function model, a *theoretical variogram* model $\gamma(\mathbf{h})$ is fitted to these values as shown in Fig. 10.4. The variogram is the expectation of squared differences of the random function $Z(x)$ of which $z(x)$ is a realisation. Based on a hypothesis of stationarity, it depends only on the separation vector **h**:

$$\gamma(\mathbf{h}) = \frac{1}{2}E\left[Z(x+\mathbf{h})-Z(x)\right]^2 \tag{10.2}$$

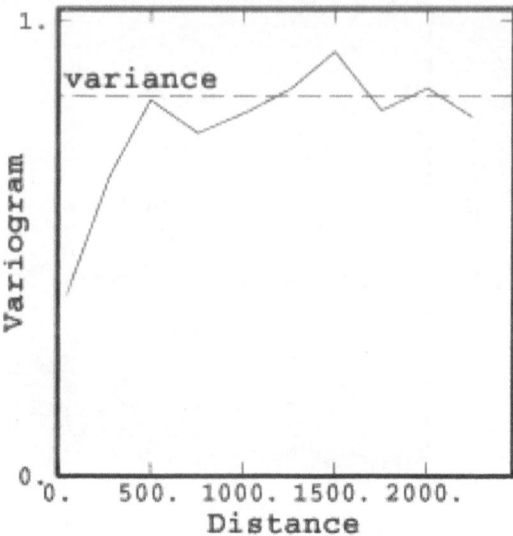

Fig. 10.3. The variogram cloud is subdivided into distance (and angle) classes. Averages are computed for each distance class, and the sequence is called the *experimental variogram*. The dashed line represents the statistical variance of the Cadmium data

The essential aspects of variogram fitting are to model the behaviour near the origin and the behaviour at large distances. The behaviour near the origin reflects

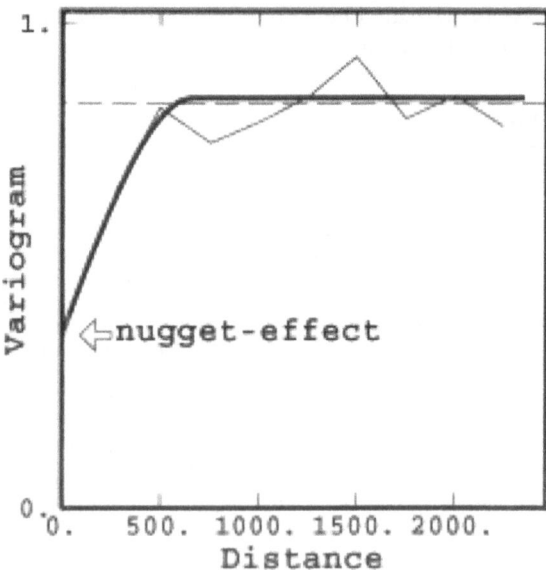

Fig. 10.4. A theoretical model is fitted to the experimental variogram. The discontinuity at the origin is called the nugget-effect

the roughness of the regionalised variable. It can be of three types:

1. **continuous and differentiable:** the variogram has a parabolic behaviour at the origin and is associated with a smooth surface;
2. **continuous but not differentiable**: the variogram has a linear behaviour at the origin and represents a rough surface;
3. **discontinuous:** the variogram shows a discontinuity at the origin and represents a discontinuous regionalised variable.

At large distances the variogram can either have an asymptotic behaviour or be unbounded. In the first case the variogram is equivalent to the classical covariance (autocorrelation) function of time series, while in the second case, the variogram represents a phenomenon which is not *second-order* stationary, but only *intrinsically* stationary (like e.g., Brownian motion). The variogram thus represents a larger class of functions for describing spatial variability than the classical covariance functions. This is why it is preferred by geostatisticians.

Common variogram models are the *nugget-effect, spherical, exponential and power* models. These models can be combined in a nested model, so that many different types of behaviour of the experimental variogram can be mimicked. Such a combination has been used in the fitting example on Fig. 10.4. A nugget-effect models the discontinuity at the origin, which can be due either to micro-scale variation or to measurement error. A spherical model with a range of 660 metres models the structured part of the data. The fact that the experimental variogram is bounded and has a sill roughly equivalent to the statistical variance (Fig. 10.3) indicates that the data can be considered as belonging to a realisation of a second-order stationary random function.

10.2.2 Kriging

The variogram model is used for constructing a multiple linear regression from data locations on a location of interest, where no measurement is available. This spatial regression has been called *kriging* after the mining engineer DG Krige, who first proposed it in the early fifties of this century.

Kriging is a transposition of multiple regression to the context of n random variables $Z(x)$ of which the data values $z(x_\alpha)$ are realisations. The regression is performed on the basis of the variogram to any arbitrary location x_α in the domain D, in order to obtain an estimated value $z*(x_o)$ (for a detailed presentation in this spirit, see Wackernagel 1998).

As an example the kriging technique is applied to a set of 259 Cadmium data from a region in the Swiss Jura (Atteia et al. 1994). Using the variogram model of Fig. 10.4, best linear unbiased estimates (BLUE) are computed at any location of the geographical domain (excluding areas where no data has been taken). The estimates on a 100 m x 100 m grid have been contoured, and the isoline map is displayed on Fig. 10.5.

At each node of the grid, a standard deviation of the estimation error of the BLUE of Cadmium has been computed and contoured. The map of the standard error for the Cadmium BLUE is shown on Fig. 10.6, on which the data locations

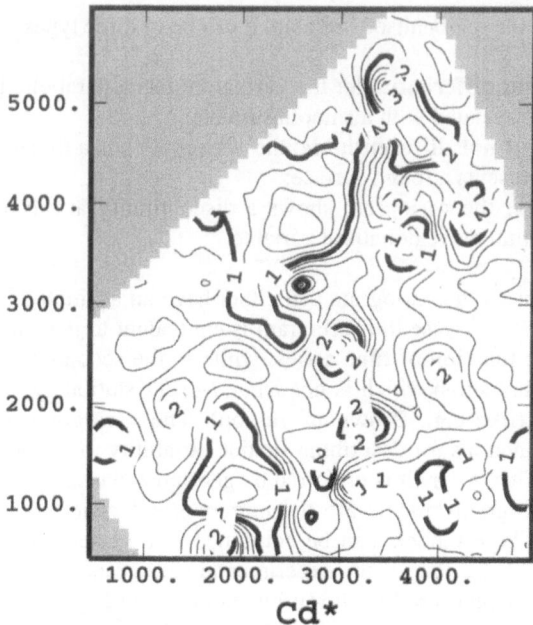

Fig. 10.5. Map of estimated Cadmium values (mg kg^{-1}) in a region of the Swiss Jura

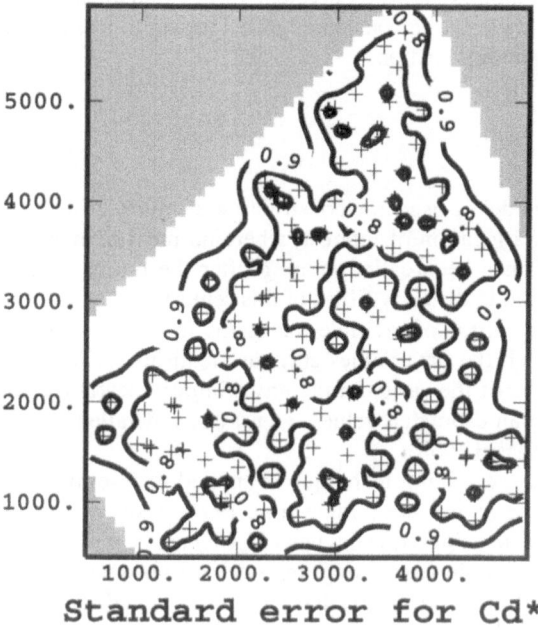

Fig. 10.6. Error map for estimated Cadmium values in the Swiss Jura. Crosses indicate data locations

have also been displayed. Because of the significant nugget-effect (micro-scale variation and measurement error), the estimation error is high in the immediate vicinity of data locations. The map of the standard error should always be delivered together with the map of the estimates, as it shows how much the latter can be trusted.

10.3 Multivariate Geostatistics

We now turn to the case when measurements of a variable of interest are supplemented by measurements of one or several auxiliary variables, eventually at different and more numerous locations. The general *cokriging* method and the more particular *external drift* method are presented. The former requires the computation of *cross-variograms* and their modeling on the basis of a *coregionalisation model*.

10.3.1 Cokriging

When data is available for different quantities Z_i ($i = 1, ..., N$), cross-variograms between pairs of variables Z_i and Z_j are defined as cross-products between differences of these variables for a pair of locations separated by a vector $\mathbf{h}$:

$$\gamma_{ij}(\mathbf{h}) = \frac{1}{2}E\left[\left(Z_i(x+\mathbf{h})-Z_i(x)\right)\cdot\left(Z_j(x+\mathbf{h})-Z_j(x)\right)\right] \qquad (10.3)$$

The theoretical cross-variogram is supposed to be translation-invariant in the random function model for a given vector $\mathbf{h}$. Experimental cross-variogram matrices for a given set of N variables can easily be fitted with nested models using a weighted least-squares algorithm described in Goulard and Voltz (1992).

The spatial regression based on the multivariate variogram model and on n data locations x_α is called *cokriging*. The estimator $Z_{i_0}^*$ for a particular variable (denoted by the index i_0) at an arbitrary location x_o is:

$$Z_{i0}^*(x_0) = \sum_{i=1}^{N}\sum_{\alpha=1}^{n_i}\omega_{i\alpha}Z_i(x_0) \qquad (10.4)$$

The weights $\omega_{i_0\alpha}^i$ for the BLUE are given by the following system of linear equations:

$$\begin{cases} \sum_{j=1}^{N}\sum_{\beta=1}^{n_j}\omega_{i_0\beta}^j\gamma_{ij}\left(\mathbf{x}_\alpha - \mathbf{x}_\beta\right)+\mu_i = \gamma_{ii_o}\left(\mathbf{x}_\alpha - \mathbf{x}_o\right) & \text{for } i=1,...\,N;\ \alpha=1,...\,n_i \\ \sum_{\beta=1}^{n_i}\omega_{i_0\beta}^i = \delta_{ii_o} & \text{for } i=1,...\,N \end{cases} \qquad (10.5)$$

In the lower part of this system, we have N conditions on the sum of the weights attached to the n_i data locations of each variable Z_i; the sums are zero for all variables except the variable of interest, and this is expressed by the Kronecker delta δ_{ii_0}. In the upper part of the system we have on the right hand the cross-variogram values for the vectors between data points $\mathbf{x}_\alpha$ and the location $\mathbf{x}_0$ as well as between all N variables and the variable of interest. On the left hand we have the cross-variograms among all variables for all pairs of data locations; we also have Lagrange parameters μ_i as a result of minimising with N constraints on the weights.

The variance of the BLUE, called the cokriging variance is computed as:

$$\sigma^2_{CKi_0} = \sum_{i=1}^{N} \sum_{\alpha=1}^{n_i} \omega^i_{i_0} \gamma_{ii_0} \left(\mathbf{x}_\alpha - \mathbf{x}_0\right) + \mu_{i_0} - \gamma_{i_0 i_0} \left(\mathbf{x}_0 - \mathbf{x}_0\right) \tag{10.6}$$

10.3.2 Data Configurations

The measurements available for different variables $Z_i(\mathbf{x})$ in a given domain may be located either at the same sample points or at different points for each variable. The following situations can be distinguished:

1. *entirely heterotopic data:* the variables have been measured on different sets of sample points and have no sample locations in common;
2. *partially heterotopic data*: some variables share some sample locations;
3. *isotopy*: data is available for each variable at all sampling points.

Entirely heterotopic data poses a problem for inferring the cross variogram or covariance model. Experimental cross variograms cannot be computed for entirely heterotopic data. Experimental cross covariances, though they can be computed, are still problematic, as the corresponding direct covariance values refer to different sets of points (and sometimes subregions). The value at the origin of the cross covariances cannot be computed.

With partially heterotopic data, it is advisable, whenever possible, to infer the cross variogram or the covariance function model on the basis of the isotopic subset of the data.

Actually, heterotopy for spatial data is as much a problem as missing values in multivariate statistics (see Example 3.1 in Wackernagel 1998), even if a model is built in – between as in the case of geostatistics.

A particular case of partial heterotopy important for cokriging is when the set of sample points of the variable of interest is included in the sets of sample points of other variables, which serve as auxiliary variables in the estimation procedure.

In this case, when the auxiliary variables are available at more points than the main variable, cokriging is typically of advantage.

In cokriging problems with heterotopic data, we can distinguish between sparsely and densely sampled auxiliary variables. In the second case, when an auxiliary variable is available everywhere in the domain, particular techniques like collocated cokriging or the external drift method can be of interest.

The question whether cokriging is still interesting in the case of isotopy (when the auxiliary variables are available at the same locations as the variable of interest) will be examined in the next section.

10.3.3 Isotopy: Intrinsic Correlation

Is the multivariate correlation structure of a set of variables independent of the spatial correlation? When the answer is positive, the multivariate correlation is said to be *intrinsic*.

A set of variables is intrinsically correlated if all variograms (direct and cross) are proportional to a basic variogram $\gamma(\mathbf{h})$:

$$\gamma_{ij}(\mathbf{h}) = b_{ij}\gamma(\mathbf{h}) \tag{10.7}$$

The matrix of coefficients b_{ij} in such an intrinsic correlation model is equivalent to the classical variance-covariance matrix of the data.

A simple way of checking for intrinsic correlation is to compute principal components and to inspect the experimental cross-variograms between the principal components: if they are zero for all lags $\mathbf{h}$, the data are compliant with the intrinsic correlation model (Wackernagel 1994).

In case of intrinsic correlation and with isotopic data, cokriging reduces to kriging of the variable of interest (Matheron 1979). All cokriging weights in Eq. 10.5 are zero for the auxiliary variables ($i \neq i_o$), when using a model as in Eq. 10.7 with isotopic data.

10.3.4 Coregionalisation Models

In general the spatial multivariate structure of the data is not intrinsically correlated, and the next simplest model, which has proven adequate in many case studies, is the linear model of coregionalisation, which is associated with nested cross-variograms:

$$\gamma_{ij}(\mathbf{h}) = \sum_{u=0}^{S} b_{uij}\gamma_u(\mathbf{h}) \tag{10.8}$$

where $\gamma_u(\mathbf{h})$ are variogram functions characterising different spatial scales of index u and b_{uij} are the variances ($i = j$) or covariances ($i \neq j$) between the variables at these scales.

A fairly recent development is the bilinear model of coregionalisation, which allows the description of systems for which the cross-covariance functions, are not even (Grzebyk and Wackernagel 1994; Wackernagel 1998).

10.3.5 Heterotopy: External Drift

An interesting case of partial heterotopy is when we have a main variable $Z(\mathbf{x})$ known at a few locations in the domain (e.g., meteorological stations) and an auxiliary variable $s(\mathbf{x})$ measured everywhere (e.g., topography).

The auxiliary variable $s(\mathbf{x})$ will enter the kriging system as a deterministic quantity (i.e., drift). The system for determining the weights ω_α attached to the values of $Z(\mathbf{x})$ at data locations $\mathbf{x}_\alpha$ is the following:

$$
\begin{cases}
\displaystyle\sum_{\beta=1}^{n} \omega_\beta \gamma\left(\mathbf{x}_\alpha - \mathbf{x}_\beta\right) + \mu_0 + \mu_1 s\left(\mathbf{x}_\alpha\right) = \gamma\left(\mathbf{x}_\alpha - \mathbf{x}_0\right) & \text{for } \alpha = 1,\ldots,n \\[2mm]
\displaystyle\sum_{\beta=1}^{n} \omega_\beta = 1 \\[2mm]
\displaystyle\sum_{\beta=1}^{n} \omega_\beta s\left(\mathbf{x}_\beta\right) = s\left(\mathbf{x}_0\right)
\end{cases}
\tag{10.9}
$$

In this system we need only the values $s(\mathbf{x}_\alpha)$ at the locations $\mathbf{x}_\alpha$ where Z is available, as well as the value $s(\mathbf{x}_0)$ at the estimation location. A successful application of the external drift method to temperature data is discussed in detail in Hudson and Wackernagel (1994).

10.3.6 Heterotopy: Collocated Cokriging

An alternative to the external drift approach is *collocated cokriging*, where we consider the auxiliary variable $S(\mathbf{x})$ as a random function (denoted by a capital "S"). We now need also the variogram of $S(\mathbf{x})$ as well as its cross-variogram with $Z(\mathbf{x})$. The cokriging system is of the form (10.5). The adjective *collocated* refers to the neighbourhood of data used, which is the same as in the external drift approach: we take for $S(\mathbf{x})$ the values co-located with the data locations of Z as well as the value co-located with the location $\mathbf{x}_0$ at which we would like to estimate Z (see Wackernagel 1998 for details). For further details see Wackernagel (1998) and Chilès and Delfiner (1999).

10.4 Non-Stationary Model

We will shortly develop the non-stationary model of geostatistics in the framework of intrinsic random functions of order k and then make the link between kriging and smoothing splines for the particular case of thin-palte splines. Splines in their statistical formulation are an interesting competitor of geostatistical methods.

10.4.1 Intrinsic Random Functions of Order k

We consider a non-stationary random function $Z(\mathbf{x})$, called an *intrinsic random function of order k* (IRF-k):

$$Z(\mathbf{x}) \;=\; Z_k(\mathbf{x}) + m_k(\mathbf{x}) \tag{10.10}$$

consisting of a deterministic part $m_k(\mathbf{x})$, representing the drift as a k-th order polynomial, and a random part $Z_k(\mathbf{x})$ with an associated generalised covariance $K(\mathbf{h})$. The polynomial of order k is a linear combination of functions $f_i(\mathbf{x})$ of the coordinates with coefficients a_l

$$m_k(\mathbf{x}) \;=\; \sum_{l=0}^{L} a_i f_k(\mathbf{x}) \tag{10.11}$$

The $L + 1$ basis functions $f_i(\mathbf{x})$ need to generate a translation invariant vector space. It can be shown that only functions belonging to the class of exponentials-polynomials fulfill this condition.

For example, in two spatial dimensions with coordinate vectors $\mathbf{x} = (\mathbf{x}_1, \mathbf{x}_2)^{\tau}$, the following monomials are often used

$$f_o(\mathbf{x}) \;=\; 1, \; f_1(\mathbf{x}) \;=\; \mathbf{x}_1, \; f_2(\mathbf{x}) \;=\; \mathbf{x}_2,$$
$$f_3(\mathbf{x}) \;=\; (\mathbf{x}_1)^2, \; f_4(\mathbf{x}) \;=\; \mathbf{x}_1 \cdot \mathbf{x}_2, \; f_5(\mathbf{x}) \;=\; (\mathbf{x}_2)^2, \tag{10.12}$$

which generate a translation invariant vector space. Trigonometric functions also belong to this class as explained in Séguret and Huchon (1990).

With estimation problems in mind we consider weights ω_α, which interpolate exactly the terms al $f_i(\mathbf{x})$ of the drift polynomial

$$\sum_{\alpha=1}^{n} \omega_\alpha a_1 f_1(\mathbf{x}_\alpha) \;=\; a_1 f_1(\mathbf{x}_0), \tag{10.13}$$

Obviously the coefficients a_l are redundant and can be dropped for the interpolation of the basis functions $f_i(\mathbf{x})$.

With weights constrained as

$$\sum_{\alpha=1}^{n} \omega_\alpha f_l(\mathbf{x}_\alpha) \;=\; f_l(\mathbf{x}_0) \quad \text{for} \quad l = 0,\dots,L \tag{10.14}$$

for any linear estimator

$$Z^*(\mathbf{x}_0) \;=\; \sum_{\alpha=1}^{n} \omega_\alpha Z(\mathbf{x}_\alpha), \tag{10.15}$$

the estimation error vanishes on average

$$E\left[Z^*(\mathbf{x}_0)-Z(\mathbf{x}_0)\right] = E\left[Z_k^*(\mathbf{x}_0)-Z_k(\mathbf{x}_0)\right] = 0 \tag{10.16}$$

By introducing a weight $\omega_0 = -1$

$$\sum_{\alpha=0}^{n}\omega_\alpha f_l(\mathbf{x}_\alpha) = 0 \quad\text{for}\quad l = 0,\dots,L \tag{10.17}$$

and we can write

$$E\left[Z^*(\mathbf{x}_0)-Z(\mathbf{x}_0)\right] = E\left[\sum_{\alpha=0}^{n}\omega_\alpha Z_k(\mathbf{x}_\alpha)\right] = 0 \tag{10.18}$$

where the weights of zero sum filter the translation-invariant drift.

The covariance function of an intrinsically stationary random function of order k is called a generalised covariance function K(h). The variance of any linear combination of values $Z_k(\mathbf{x}_\alpha)$ with weights constrained to

$$\sum_{\alpha=0}^{n}\omega_\alpha f_l(\mathbf{x}_\alpha) = 0 \quad\text{for}\quad l = 0,\dots,L \tag{10.19}$$

equals

$$\begin{aligned}
\operatorname{var}\left(\sum_{\alpha=0}^{n}\omega_\alpha Z_k(\mathbf{x}_\alpha)\right) &= E\left[\left(\sum_{\alpha=0}^{n}\omega_\alpha Z_k(\mathbf{x}_\alpha)\right)^2\right]\\
&= \sum_{\alpha=0}^{n}\sum_{\beta=0}^{n}\omega_\alpha\omega_\beta K(\mathbf{x}_\alpha-\mathbf{x}_\beta)
\end{aligned} \tag{10.20}$$

A generalised covariance is a *k-th order conditionally positive definite* function

$$\sum_{\alpha=0}^{n}\sum_{\beta=0}^{n}\omega_\alpha\omega_\beta K(\mathbf{x}_\alpha-\mathbf{x}_\beta) \geq 0$$
$$\text{for}\quad \sum_{\alpha=0}^{n}\omega_\alpha f_l(\mathbf{x}_\alpha) = 0, \quad l = 0,\dots L \tag{10.21}$$

From this definition we see that the negative of the variogram, $-\gamma(\mathbf{h})$, is the generalised covariance of an intrinsic random function of order zero. Increasing the order k puts more restrictions on the weights but enlarges the class of possible covariances. The theory of intrinsically stationary random functions of order provides a generalisation of the random functions with second order stationary increments, which themselves were set in a broader framework than the second order stationary functions.

The inference of the highest order of the drift polynomial to be filtered is usually performed using automatic algorithms, a procedure which we shall not describe here (see Chiles and Delfiner 1999, for details).

A popular generalised covariance is the polynomial model

$$K_{pol}(\mathbf{h}) = \sum_{u=0}^{k} b_u (-1)^{u+1} |\mathbf{h}|^{2u+1} \quad \text{with} \quad b_u \geq 0 \tag{10.22}$$

The conditions on the coefficients b_u are sufficient. Looser bounds on these coefficients are given in Matheron (1973).

The polynomial generalised covariance model is a nested model, which has the striking property that it is built up with several structures having a different behaviour at the origin. For linear drift ($k = 1$), the term for $u = 0$ is linear at the origin and is adequate for the description of a regionalised variable, which is continuous but not differentiable. The term of the polynomial generalised covariance for $u = 1$ is cubic and thus appropriate for differentiable regionalised variables.

If a nugget-effect component is added to the polynomial generalised covariance model, discontinuous phenomena are also covered. This extended polynomial generalised covariance model is flexible with respect to the behaviour at the origin, and well suited for automatic fitting.

10.4.2 Kriging with Drift

For kriging a non-stationary random function $Z(\mathbf{x})$ at an arbitrary location $\mathbf{x}_0$ of the spatial domain we take a linear combination of weights ω_α with data at locations $\mathbf{x}_\alpha$:

$$Z^*(\mathbf{x}_0) = \sum_{\alpha=1}^{n} \omega_\alpha Z(\mathbf{x}_\alpha) \tag{10.23}$$

We want no bias

$$E\left[Z(\mathbf{x}_0) - Z^*(\mathbf{x}_\alpha)\right] = 0 \tag{10.24}$$

which yields

$$m(\mathbf{x}_0) - \sum_{\alpha=1}^{n} \omega_\alpha m(\mathbf{x}_\alpha) = 0 \tag{10.25}$$

and

$$\sum_{l=0}^{L} a_l \left(f_l(\mathbf{x}_0) - \sum_{\alpha=1}^{n} \omega_\alpha f_l(\mathbf{x}_\alpha) \right) = 0 \tag{10.26}$$

As the a_l are nonzero, the following set of constraints on the weights ω_α emerges

$$\sum_{l=0}^{L} \omega_\alpha f_l(\mathbf{x}_\alpha) = f_l(\mathbf{x}_\alpha) \quad \text{for} \quad l = 0,\dots,L \tag{10.27}$$

For the constant function $f_o(\mathbf{x})$, this is the usual condition

$$\sum_{\alpha=1}^{n} \omega_\alpha = 1 \tag{10.28}$$

Developing the expression for the estimation variance, introducing the constraints into the objective function together with Lagrange parameters μ_l, and minimising, we obtain the kriging system

$$\begin{cases} \sum_{\beta=1}^{n} \omega_\beta K(\mathbf{x}_\alpha - \mathbf{x}_\beta) - \sum_{l=0}^{L} \mu_l f_l(\mathbf{x}_\alpha) = K(\mathbf{x}_\alpha - \mathbf{x}_0) & \text{for } \alpha = 1,\dots,n \\[2mm] \sum_{\beta=1}^{n} \omega_\beta f_l(\mathbf{x}_\beta) = f_l(\mathbf{x}_0) & \text{for } l = 0,\dots,L \end{cases} \tag{10.29}$$

and in matrix notation

$$\begin{pmatrix} \mathbf{K} & \mathbf{F} \\ \mathbf{F}^\tau & 0 \end{pmatrix} \begin{pmatrix} \mathbf{w} \\ -\mu \end{pmatrix} = \begin{pmatrix} \mathbf{k} \\ \mathbf{f} \end{pmatrix}, \tag{10.30}$$

where $\mathbf{K}$ is the matrix of generalised covariances $K(\mathbf{x}_\alpha - \mathbf{x}_\beta)$ between data locations, while k is the vector of generalised covariances $K(\mathbf{x}_\alpha - \mathbf{x}_0)$ between data locations and the estimation point.

For this system to have a solution, it is necessary that the matrix

$$\mathbf{F} = (\mathbf{f}_0,\dots,\mathbf{f}_L) \tag{10.31}$$

is of full column rank, i.e., the column vectors $\mathbf{f}_l$ have to be linearly independent. This means in particular that there can be no other constant vector besides the vector $\mathbf{f}_0$. Thus, the functions $f_l(\mathbf{x})$ have to be selected with care.

10.4.3 Dual Kriging

We now examine the question of defining an interpolation function based on the kriging system. The interpolator is the product of the vector of samples $\mathbf{z}$ with a vector of weights $\mathbf{w}_{\mathbf{x}}$, which depends on the location in the domain

$$z^*(\mathbf{x}) = \mathbf{z}^\tau \mathbf{w}_{\mathbf{x}} \tag{10.32}$$

The weight vector is solution of the kriging system

$$\begin{pmatrix} \mathbf{K} & \mathbf{F} \\ \mathbf{F}^\tau & \mathbf{0} \end{pmatrix} \begin{pmatrix} \mathbf{w}_{\mathbf{x}} \\ -\mu_{\mathbf{x}} \end{pmatrix} = \begin{pmatrix} \mathbf{k}_{\mathbf{x}} \\ \mathbf{f}_{\mathbf{x}} \end{pmatrix} \tag{10.33}$$

in which all terms dependent on the estimation location have been subscripted with an $\mathbf{x}$. In this formulation we need to solve the system each time for each new interpolation location. As the left hand matrix does not depend on $\mathbf{x}$, let us define its inverse (assuming its existence) as

$$\begin{pmatrix} \mathbf{T} & \mathbf{U} \\ \mathbf{U}^\tau & \mathbf{V} \end{pmatrix} \tag{10.34}$$

The kriging system is

$$\begin{pmatrix} \mathbf{w}_{\mathbf{x}} \\ -\mu_{\mathbf{x}} \end{pmatrix} = \begin{pmatrix} \mathbf{T} & \mathbf{U} \\ \mathbf{U}^\tau & \mathbf{V} \end{pmatrix} \begin{pmatrix} \mathbf{k}_{\mathbf{x}} \\ \mathbf{f}_{\mathbf{x}} \end{pmatrix} \tag{10.35}$$

The interpolator can thus be written

$$z^*(\mathbf{x}) = \mathbf{z}^\tau \mathbf{T} \mathbf{k}_{\mathbf{x}} + \mathbf{z}^\tau \mathbf{U} \mathbf{f}_{\mathbf{x}} \tag{10.36}$$

Defining $\mathbf{b}^\tau = \mathbf{z}^\tau \mathbf{T}$ and $\mathbf{d}^\tau = \mathbf{z}^\tau \mathbf{U}$, the interpolator is a function of the right hand side of the kriging system

$$z^*(\mathbf{x}) = \mathbf{b}^\tau \mathbf{k}_{\mathbf{x}} + \mathbf{d}^\tau \mathbf{f}_{\mathbf{x}} \tag{10.37}$$

Contrary to the weights $\mathbf{w}_{\mathbf{x}}$, the weights $\mathbf{b}$ and $\mathbf{d}$ do not depend on the target point $\mathbf{x}$.

Combining the data vector $\mathbf{z}$ with a vector of zeroes, we can set up the system

$$\begin{pmatrix} \mathbf{T} & \mathbf{U} \\ \mathbf{U}^\tau & \mathbf{V} \end{pmatrix} \begin{pmatrix} \mathbf{z} \\ \mathbf{0} \end{pmatrix} = \begin{pmatrix} \mathbf{b} \\ \mathbf{d} \end{pmatrix} \tag{10.38}$$

which, once inverted, yields the *dual system* of kriging

$$\begin{pmatrix} \mathbf{K} & \mathbf{F} \\ \mathbf{F}^\tau & \mathbf{0} \end{pmatrix} \begin{pmatrix} \mathbf{b} \\ \mathbf{d} \end{pmatrix} = \begin{pmatrix} \mathbf{z} \\ \mathbf{0} \end{pmatrix} \tag{10.39}$$

There is no reference to any interpolation point $\mathbf{x}$ in this system: it needs to be solved only once for a given region.

It should be noted that when the variable to investigate is equal to one of the deterministic functions, e.g. $z(\mathbf{x}) = f_1(\mathbf{x})$, the interpolator is $z^*(\mathbf{x}) = \mathbf{w}^\tau \mathbf{f}_1$. As the weights are constrained to satisfy $\mathbf{w}^\tau \mathbf{f}_l = f_l(\mathbf{x})$ we see that $z^*(\mathbf{x}) = f_l(\mathbf{x}) = z(\mathbf{x})$. Thus the interpolator is exact for $f_l(\mathbf{x})$. This clarifies the meaning of the constraints in kriging: the resulting weights are able to interpolate exactly each one of the deterministic functions.

10.4.4 Splines

The mathematical spline took its name from the draftmen's mechanical spline, which is "a thin reedlike strip that was used to draw curves needed in the fabrication of *cross*-sections of ships' hulls" (Wahba 1990b). We shall restrain discussion to the case of smoothing thin-plate splines. A general framework for splines (with an annotated bibliography) is presented in Champion et al. (1996). The equivalence between splines and kriging is analysed in more depth in Matheron (1981) and Wahba (1990b).

The model for the smoothing splines can be written as

$$Z(\mathbf{x}) = \underbrace{\Phi(\mathbf{x})}_{\text{smooth}} + \underbrace{Y(\mathbf{x})}_{\text{white noise}} \tag{10.40}$$

where $\Phi(\mathbf{x})$ is uncorrelated with the white noise (i.e. nugget-effect) component $Y(\mathbf{x})$. The term $\Phi(\mathbf{x})$ is estimated by a smooth function g minimising

$$\sum_{\alpha=1}^{n} \left(\Phi(\mathbf{x}_\alpha) - g(\mathbf{x}_\alpha) \right)^2 + \lambda J_p(g) \tag{10.41}$$

with $J_p(g)$ a measure of roughness (in terms of pth degree derivatives) and $\lambda > 0$ a smoothing parameter.

The $g(\mathbf{x})$ function is written as the sum of two terms

$$g(\mathbf{x}) = \sum_{\alpha=1}^{n} b_\alpha \psi(\mathbf{x} - \mathbf{x}_\alpha) + \sum_{l=0}^{L} d_l f_l(\mathbf{x}) \tag{10.42}$$

where

$$\psi(\mathbf{x} - \mathbf{x}_\alpha) = K(\mathbf{x} - \mathbf{x}_\alpha) = |\mathbf{h}|^2 \log|\mathbf{h}| \tag{10.43}$$

is a $p-1$ conditional positive definite function (in geostatistics this generalised covariance is known as the *spline covariance* model).

The weights b_α and d_l are the solution of

$$\begin{pmatrix} \mathbf{K} + \lambda\mathbf{I} & \mathbf{F} \\ \mathbf{F}^\tau & 0 \end{pmatrix} \begin{pmatrix} \mathbf{b} \\ \mathbf{d} \end{pmatrix} = \begin{pmatrix} \mathbf{z} \\ \mathbf{0} \end{pmatrix} \tag{10.44}$$

This system is equivalent to the dual kriging system (10.39). In the random function model we can understand the term $\lambda\mathbf{I}$ as a nugget-effect (white noise) added to the variances at data locations, but not to the variance at the estimation location.

So, in geostatistical terms, the system (10.44) represents a filtering of the nugget-effect component on the basis of a non-stationary linear model of regionalisation (see Wackernagel 1998, p 119).

In the spline approach, the parameters are obtained by *generalised cross validation* (GCV), which is a predictive mean square error criterion (leave-one-out technique) to estimate the degree $p = k+1$ of the spline and the smoothing parameter λ.

Synthetic examples are discussed in Wahba (1990) on pages 46 – 47 and 48 – 50. Dubrule (1984) presents an example from oil exploration from the point of view of geostatistics. Hutchinson and Gessler (1994) have treated the same data set with splines and show that they can obtain equivalent results; in particular, they provide prediction errors from a Bayesian model, which are analogous to the kriging standard deviations.

Kriging is usually considered in the atmospheric sciences as a variant of objective analysis. The equivalence between objective analysis and splines is discussed in detail in Bennett (1992), with examples from oceanography, using traditional covariance functions instead of generalised covariance functions.

10.5 Conclusion

We have introduced only a few concepts of linear geostatistics in this paper. Other techniques of interest are geostatistical simulations (Armstrong and Dowd 1994) or non-linear geostatistics (Rivoirard 1994). A very complete reference text on all aspects of geostatistics has been published by Chilès and Delfiner (1999). Geostatistical space-time models are reviewed in Kyriakidis and Journel (1999).

Concerning the relation between splines and kriging, we have seen that

- geostatistics models explicitly the autocorrelation K(h) with respect to a polynomial drift of order k;
- splines infer the smoothness parameter λ and the degree $p = k+1$ by generalised cross-validation;
- smoothing splines are equivalent to kriging with a translation-invariant drift filtering noise.

The analysis of the relations between both approaches can provide many useful insights, especially in a more abstract framework of presentation than the one adopted here. With a view on statistical data analysis let us conclude like Cressie (1991), stating that our "preference is with kriging's obligatory spatial-dependence assessment and its automatic calculation of mean-squared prediction errors."

Chapter 11
Statistics – an Indispensable Tool in Dynamical Modeling

by Hans von Storch

Abstract

The role of statistical analysis in the process of establishing and utilising ocean and other environmental models is discussed. A general state space model approach is adopted. In "quasi-realistic models," statistical thinking is encoded in the parameterisations and is required for extracting experimental evidence and for validation. Data assimilation techniques are used to systematically combine observational evidence and quasi-realistic models. While quasi-realistic models serve as complex substitute reality, is dynamical knowledge represented through simplified models. These "cognitive" idealised models have to be fitted to observational data when adapted to real situations.

11.1 Environmental Research

As outlined in a lecture of this school (von Storch 2000), two fundamentally different types of mathematical models are used in environmental research. One sort is "quasi-realistic," and is supposed to be a substitute reality, within which the otherwise impossible experiments can be conducted. A few decades earlier, such models were often mechanistic, but most of these apparata have now been replaced by mathematical models (Sündermann and Vollmers 1972). They are also used to extra- and interpolate in a dynamically consistent manner the sparse observations, so that spatially and temporally high resolution analyses of the system's state are constructed. A representative of this type are 3-D models of the North Sea with a resolution of a few tens of kilometers, simulating explicitly an array of processes such as advection, mixing, tides, bottom stress, wind stress, air-sea interaction and the like (e.g., Kauker 1999). The other type of model, named here "cognitive," is highly simplified and idealised. *Because of its reduced complexity,* such a model constitutes "knowledge." An example of this type of model is Frankignoul's model of the variability of the mixed layer depth (Frankignoul 1979). Both types of modeling require the use of statistical thinking, both in terms of design as well as analysis.

The present paper is a discussion about the different roles of statistics in designing and using these models. Throughout this discussion, we make use of the formalism of *state space models* (Sect. 11.2). In Sect. 11.3, the role of statistics in quasi-realistic modeling is considered: parameterisations, forecasting, simulation and numerical experimentation, and data analysis. Several examples, ranging from the specification of high frequency wind fluctuations used in wave modeling, hindcasting storm surge statistics over 40 years, and simulating the impact of increased atmospheric carbon dioxide concentrations on storm surge statistics are presented. In Sect. 11.3 the role of statistics for "cognitive" models is discussed; first the general concept of *Principal Interaction Patterns* is introduced, and an example of a *Principal Oscillation Pattern* Model of wave dynamics along the Pacific equator is presented.

11.2 State Space Models

Here, we introduce as a kind of overarching general view the concept of *state space models*.

We describe our system with a state variable, represented by an m-dimensional state vector ϕ_t. Often, this state vector can not be observed, sometimes because of lack of suitable sensors, but also because space-time continuous observations are not doable. The dynamics of this state variable are described by a system of difference equations with the dynamics F.

$$\phi_{t+1} = F\!\left(\phi_t, \vec{\alpha}, t\right) + \varepsilon \tag{11.1}$$

The dynamics depend on a set of free parameters $\vec{\alpha} = (\alpha_1, \alpha_2, \ldots)$; the fact that the dynamics are only approximately known as well as the fact that seemingly random effects act upon our system is taken care of by ε. Of course, the dynamics may generate internal noise as well. The dynamics F may be derived from theoretical arguments, such as the conservation of momentum or mass, or after an empirical fit.

Even if ϕ_t is not observable in its entirety, some empirical evidence will be available, for instance at some locations and at some times, or as indirect evidence from proxy information such as lake waves. If these observations are combined into an observation vector ω_t, we have an *observation equation* that relates the state variable to the observed variables

$$\omega_t = P(\phi_t) + \delta \tag{11.2}$$

Again, the observation equation is not exactly satisfied; there may be measurement uncertainties with respect to the value or to location and timing; also the link may be a bit fuzzy as in case of proxy data. The operator P is usually not invertible.

11.3 Statistics and Quasi-realistic Models

In this section we deal with quasi-realistic models, i.e., numerical models that incorporate all processes of first, second and sometimes third order. The level of complexity of such models is usually limited by the available computer power.

We discuss the problem of parameterisations, of analysing the output from quasi-realistic models and the problem of how to estimate the trajectory of an open system with the help of a quasi-realistic model and observational data.

11.3.1 Parameterisations

In the design of quasi-realistic models, such as GCM-type global or regional ocean models, the unavoidable truncation of the basic dynamical equations, necessitates the introduction of parameterisations of sub-grid scale processes (cf. von Storch et al. 1999). Such processes, such as the turbulent layers of the ocean at the surface and at the bottom, take place on scales too small to be resolved, but have a significant impact on the dynamics of the state variables such as the stream function. Therefore their *effect* on the resolved scales is considered to be partially determined by the configuration given at grid size scale (von Storch 1997). In *the state space formalism*, the fitting of parameterisations means to specify some of the unknown parameters α.

To allow for formalisation, let us write the state variable ϕ as a sum of the large-scale resolved component ϕ and an unresolved part ϕ'

$$\phi = \bar{\phi} + \phi' \tag{11.3}$$

Then, our basic differential equations

$$\frac{\partial \phi}{\partial t} = F(\phi) \tag{11.4}$$

with the "dynamics" F is replaced by

$$\frac{\partial \bar{\phi}}{\partial t} = F_{\Delta x}(\phi) \tag{11.5}$$

with a modified operator $F_{\Delta x}$ resulting from the full operator F after introducing a truncated spatial resolution Δx. In general, this operator may be written as

$$F_{\Delta x}(\phi) = F(\bar{\phi}) + F'(\phi') \tag{11.6}$$

with an operator F' describing the net effect of the sub grid scale variations represented by ϕ'. With this set-up, the system (5) is no longer closed and can therefore no longer be integrated. For overcoming this problem, conventional approaches assume that the "nuisance" term $F'(\phi')$ is either irrelevant, i.e.,

$$F'(\phi') = 0 \tag{11.7}$$

or may be parameterised by

$$F'(\phi') = Q(\bar{\phi}) \tag{11.8}$$

with a dynamically motivated function Q. For the specification of this function, usually a number of parameters are to be determined. This is mostly done by statistically fitting Eq. 11.8 to data from observational campaigns.

While both specifications (11.7 – 8) return an integrable Eq. 11.5, they both have to assume that the local scale acts as a deterministic slave of the resolved scales. However, in reality there is variability at local scales *unrelated to the resolved scales*. Thus, Eq. 11.5 should take into account that $F'(\phi')$ can not completely be specified as a function of ϕ, but that formulation (11.5) should be replaced by

$$F'(\phi') \sim S(\vec{\alpha}) \tag{11.9}$$

with a random process S with parameters $\vec{\alpha}$ which are conditioned upon the resolved state ϕ:

$$F'(\phi') \sim S\left(R(\bar{\phi})\right) \tag{11.10}$$

When the mean value $\mu(\phi)$ is the only parameter in the vector $\vec{\alpha}$, which depends on ϕ, then the distribution S may be written as

$$S\left(R(\bar{\phi})\right) = \mu(\bar{\phi}) + S' \tag{11.11}$$

with a conditional mean value $\mu(\phi)$ and a random components with zero mean value $(E(S') = 0)$ and uncertainty unrelated to the resolved scales. Specification (11.8) equals specification (11.11) if $S' = 0$ and $\mu(\bar{\phi}) = Q(\bar{\phi})$.

The role of statistics here is to first suggest a suitable distribution S, and later to condition the free parameters $\vec{\alpha}$ in a manner such that observed or otherwise physically meaningful relationships are approximately satisfied. These parameters often include parameters such as means, variances and lag correlations or spectra.

A comparison of a randomised parameterisation with a conventional parameterisation has been provided by Bauer and Weisse (2000), who ran the ocean wave model WAM for an extended period of 6 months with analysed winds. These winds are available only every 6 hours. Usually the variability within these 6 hours, related to various high-frequency meteorological events such as passage of fronts and the associated gustiness is disregarded, and the wind is either kept constant or, as in the present case, linearly interpolated.

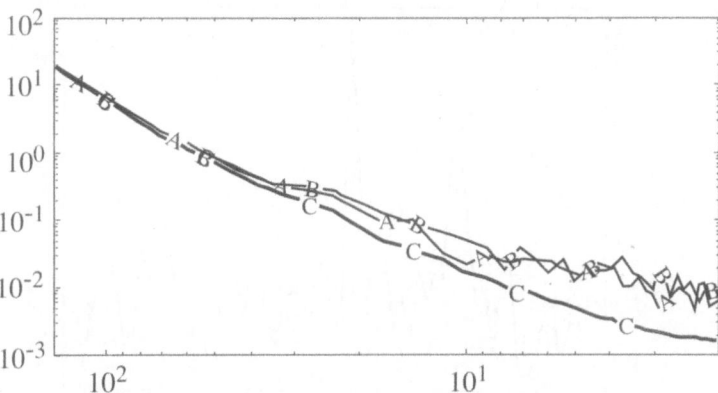

Fig. 11.1. Spectra of wind speeds at Ekofisk; time resolution 20 min (i.e., the abscissa of 9 corresponds to 9 x 20 min = 3 hours), units: (m/s) 2 x 20 min. A is the spectrum of the original series, B is for the 6 hourly wind speed with randomised interpolation and C with linear interpolation.
From Weisse and Bauer (pers. comm.)

In their study, Bauer and Weisse considered the linear interpolation as the "conditional expectation" parameterisation (11.8) of short term atmospheric fluctuations. A randomised version was constructed from time series of wind observations every 20 minutes. It returns for any instantaneous pair of wind speeds 6 hours apart a consistent series of 20 minutes wind speeds. This constitutes their "randomised parameterisation" (11.11). The success of the random number generator is demonstrated in Fig. 11.1, displaying three auto spectra. One spectrum (A) is from the original series observed every 20 min. In the other two spectra, first a wind speed was sampled every 6 hours, and then either linearly interpolated (C) or random numbers were inserted (B). The spectra are very similar for time scales longer than 6 hours (corresponding to an abscissa of 18); the linear interpolation causes a severe underestimation of the high-frequency variance, which is entirely recovered by the randomised parameterisation.

The effect of the two different formulations of high-frequency wind variations on the statistics of ocean waves was analysed in terms of the distribution of wave heights every 20 minutes within 3 days. For each interval of three days, the 6th largest wave height is determined, i.e., the 97% from the sample of 216 wave height values within three days. The time series of these 97% quantiles for 60 consecutive 3-day intervals for the two WAM simulations is displayed in Fig. 11.2. Obviously, the two time series are very similar, so that the presence of short term fluctuations does not induce significant low-frequency variations; thus, the randomised parameterisation is not required for the simulation of the overall statistics. However, a closer inspection reveals that the distributions are shifted to taller waves (not only in terms of 97% quantiles but also 50% quantiles); that is, the presence of high-frequency variations may increase the height of extreme values by a few decimeters.

It seems that while a "randomised parameterisation" (11.9) is theoretically attractive, in most applications the additional introduction of variability is inconse-

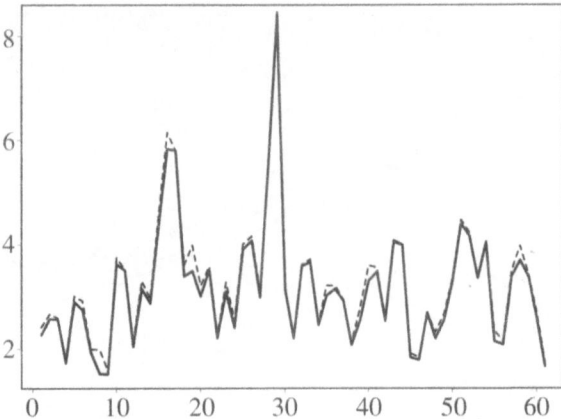

Fig. 11.2. Time series of 97 % quantiles of significant wave height at 30°N, 30°W calculated for consecutive 3-day intervals. The solid line is from the run with linearly interpolated winds, and the dashed curve from the simulation with a randomised specification. Units: m/s. Abscissa: number of 3-day long intervals.
After Bauer and Weisse (2000)

quential, because the dissipative mechanisms of the models efficiently remove variability on the smallest scales. However, experiments with an Energy Balance Model (von Storch et al. 1999), an idealised ocean model (Timmermann 1999) and high-frequency wind forcing of a wave model (Bauer and Weisse 2000) demonstrate the potential of this approach.

11.3.2 Analyzing Integrations of Quasi-Realistic Models

When quasi-realistic models are integrated, the purpose may be forecasting, simulation or executing a numerical experiment.

11.3.2.1 Forecasting

Forecasts, like the prediction of stream flows in rivers, are made to provide users with timely information to allow for adaptation to a changing environment. This information is usually a point-value, such as a river level at a given time, or an interval, such as a temperature range, or a probability, such as the probability for rain on a given day. The information provided by a single forecast is easy to grasp, but the information provided of the forecast system needs a *statistical analysis* of the predictive skill, concerning the frequency of hits or the expected error (Murphy and Epstein 1989). Sometimes efforts are made to add to the prediction of state variables a prediction of the skill of the forecast itself.

In terms of the *state space model*, forecasting means to first determine "initial states" by suitably solving the observational Eq. 11.2 and then integrating the state space Eq. 11.1 forward in time. From the large-scale forecast, local forecasts are derived from "model output statistics" (Klein and Lahn 1974), which amount to invoking another observation model. In most cases, the uncertainty terms ε and δ are disregarded, but not always.

11.3.2.2 Simulations

Simulations are made to generate a quasi-realistic extended trajectory of the considered system. Often, it is simply impossible to observe with sufficient spatial and temporal resolution the development of the system. For instance, a high resolution current field of the North Sea can not determined from observations, let alone for an extended time. Then, it is advisable to run a quasi-realistic model instead. However, the result of this model is highly complex, not as complex as reality, but in practical terms much too complex to grasp the wide range of phenomena and their dynamics. Thus, the evaluation of a simulation requires the skillful *statistical analysis* of the output of such a model. Without such an analysis, the researcher can only consider a limited number of events, analyse these events in terms of the processes involved, and compare them with observations.

An important aspect with such simulations is the validation of the models. There are cases where models have shown sensitivities, mainly because the models replicate not the real system but somewhat different, overly sensitive systems.

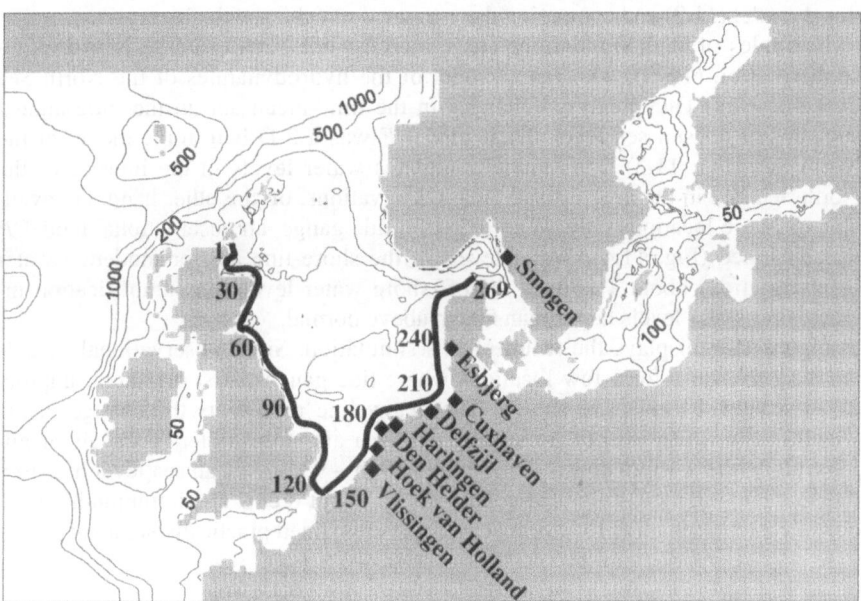

Fig. 11.3. Integration area in Langenberg's study. The black line indicates the location of the considered 270 near-coastal water level variations
From Langenberg et al. (1999)

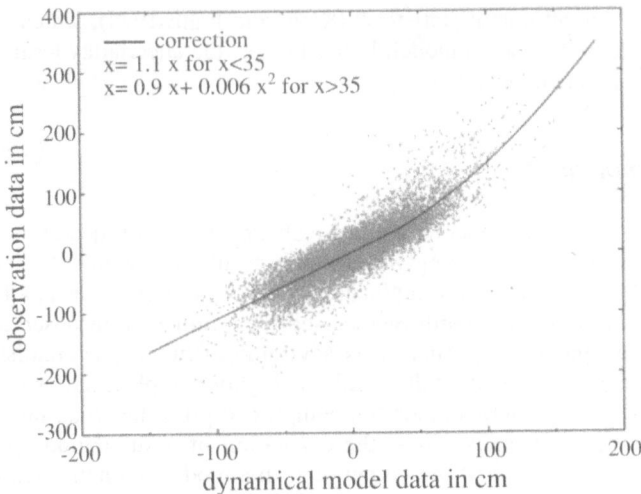

Fig. 11.4. Observations model, relating near coastal water level (horizontal axis) to shore-line water level (vertical axis), for Langenberg's 2-D North Sea model

In terms of the *state-space model*, simulations are done by integrating Eq. 11.1, mostly by disregarding the ε-term. Confirmation is done by invoking an observational model (11.2) and comparing the estimated ω-values with observations.

Examples of such simulations are numerous; one such example is from Langenberg et al. (1999), who ran a model of the hydrodynamics of the North Sea over 40 years. They analysed changes in the time-mean and in the intra-annual statistics of coastal sea level. Their model F was a 2-D barotropic model of the North Sea (Fig. 11.3). This model simulates water levels in the interior of the North Sea and in the near-coastal area. Observations, on the other hand, are available on shore locations only. With help of a tide gauge, an observational model P was designed (Fig. 11.4). Water levels at the shore-line are for moderate water level variations about 1.1 times the off-shore water levels; the amplification increases for water levels larger than 50 cm above normal.

Figure 11.5 displays the result of this simulation. Since observational models are available only for a few locations where tide gauges exist off-shore, unprocessed sea level trends are shown. The heavy line represents the change in the winter mean, which is of the order of 2 mm/year along the eastern coast and negligible along the western coast. Thus, within 1955 – 93, the mean sea level has risen due to changing meteorological conditions by about 8 cm. The intramonthly 90% quantiles, representative for storm activity, has remained practically stationary.

11.3.2.3 Numerical Experimentation

Quasi-realistic models, which have been tested in simulations of the historical record, may also be used to test the system's sensitivity to changes of boundary conditions, forcing functions and internal processes. One simulation is done with

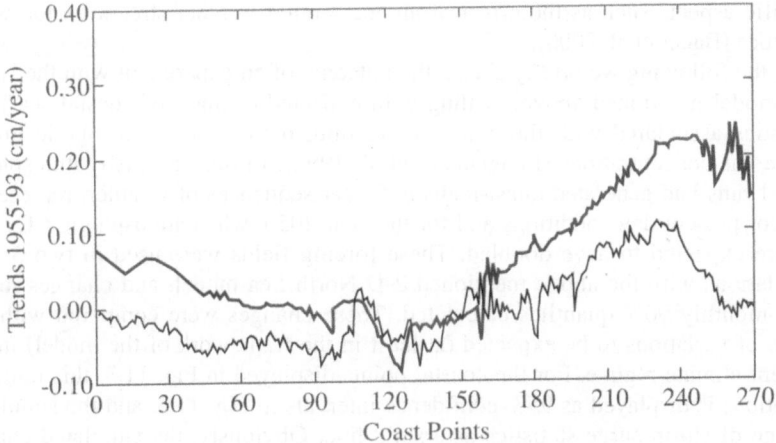

Fig. 11.5. Trends of winter means (heavy line) and winter 90% quantiles (light line) of winter high tide levels for 270 near-coastal locations along the North Sea coast (as indicated in Fig. 11.3) simulated in a 1955 – 93 model integration. Units: cm/year. From Langenberg et al. (1999)

"control" conditions, and others with a limited number of factors modified in a perfectly controlled manner. In this way, the effect of the "treatment" may be examined. Like in medical research, a *statistical analysis* of the outcome of such an experiment is often required because of the inherently noisy character of environmental systems. Usually, decision rules are required to reject null hypotheses, "treatment has no effect," with a given risk.

These "treatments" may be dramatic as the effect of opening the Central American Isthmus in Panama (Maier-Reimer et al. 1990), but may also feature different parameterisations of the cirrus clouds (Lohmann and Roeckner 1995), or

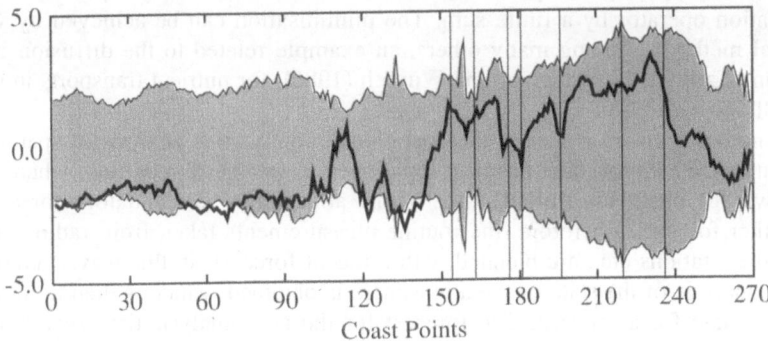

Fig. 11.6. Change of winter 90% quantiles of winter high tide levels for 270 near-coastal locations along the North Sea coast (as indicated in Fig. 11.3) from today to the time of expected doubling of atmospheric CO_2 concentration, in cm. Shaded area: 95% confidence interval for present natural variability; heavy line: expected anthropogenic change. From Langenberg et al. (1999)

specific aspects such as the effect of an accelerated weather stream on the wave statistics (Bauer et al. 2000).

In the following we briefly sketch the outcome of an experiment with the North Sea model mentioned above, dealing with expected changes of coastal sea level statistics associated with the expected ongoing increase of atmospheric greenhouse gas concentrations (Langenberg et al. 1999). In this case, two atmospheric model runs had generated quasi-realistic 5-year sequences of weather, representative for present day conditions and for the year 2050, when atmospheric CO_2 levels are expected to have doubled. These forcing fields were used in two 5-year simulations with the above mentioned 2-D North Sea model, and changes in the intra-monthly 90% quantiles calculated. These changes were compared with the range of variations to be expected (at least in the framework of the model) in the present climate regime. For the coastal points displayed in Fig. 11.3, this range of variations is displayed as 95% confidence intervals in Fig. 11.6, and the simulated change of storm surge statistics as a solid line. Obviously, the simulated change varies within the 95% confidence band, indicating that the treatment "modified atmospheric composition" has little effect on the statistics of storm surges along the North Sea coast.[1]

11.3.3 Merging Dynamical Knowledge and Observational Evidence

A standard exercise is the determination of parameters $\vec{\alpha}$ in the state space Eq. 11.1. In that case, the observational evidence ω is usually a series of statistics, such as spatial variances, and the parameters are determined such that

$$E\left(\left\|\omega_{t+1} - P\big(F(\phi_t, \vec{\alpha}, t)\big)\right\|\right) = \min, \tag{11.12}$$

where $E(\cdot)$ represents the expectation operator. In practical cases, only a limited number of observations is available, and the usual statistical assumptions about stationarity and ergodicity have to be made before reasonably replacing the expectation operator by a finite sum. The minimisation can be achieved by variational methods. Among many others, an example related to the diffusion in the ocean is provided by Schröter and Wunsch (1986), for nutrient transports in rivers by Bülow et al. (1998).

Another problem is that of the operational, consecutive analysis of spatial distributions. This procedure has been developed in weather forecasting, where good knowledge about the (initial) state of the atmosphere is mandatory for a good weather forecast. Therefore, the routine measurements taken from radio sondes, weather stations etc., are blended with a recent forecast. In this way, a complete 3–D analysis of the state of the atmosphere is obtained, which is then used as the initial state for a forecast. The forecast for the next analysis time serves as the "first guess" to be blended with the new observations.

---

[1] Note that the effect of an overall increase of the ocean's volume, due to thermal expansion and changes of the mass of glaciers and ice sheets is not included in this analysis.

A large number of different techniques have been developed in the past; an overview, illustrated with examples, is provided by Robinson et al. (1998). The above sketched consecutive approach can be seen as a kind of *Kalman filter* (e.g., Honerkamp 1994, or Jones 1985) with a state model (11.1) and an observation model (11.2). With the state model a forecast

$$\phi_t^f = F\left(\phi_{t-1}, \vec{a}, t-1\right) \tag{11.13}$$

is prepared, from which consistent observations ω_t^f are estimated with the help of the observation model:

$$\omega_t^f = P\left(\phi_t^f\right) \tag{11.14}$$

These estimated values are compared with the actual observations ω_t and a final "best estimate" ϕ_t is determined as linear combination

$$\phi_t = \phi_t^f + K\left(\omega_t - \omega_t^f\right) \tag{11.15}$$

with a suitable operator K, which depends on the covariance matrices of the noise terms ε and δ in Eq. 11.1 – 2.

11.4 Reduced "Cognitive" Models

As already discussed, are quasi-realistic models are too complex to allow for *understanding*, in much the same way that simply looking at environmental phenomena without a-priori theoretical reasoning is rarely a means of understanding the underlying mechanisms. Therefore, certain conceptual "models" of the phenomena are usually set up and data are checked to the extent they are consistent with the conceptual model. The conceptual model is an idealised description of the real situation, stripped down to a few relevant processes and their interaction. When supported by empirical evidence, the conceptual model embodies "understanding."

"Conceptual model" means that the functional from of dependencies is fixed, but that certain parameters are left to be fitted to the data. They may formally expressed in terms of *Principal Interaction Patterns*, as introduced by Hasselmann (1988). The concept has been implemented fully only in a few cases (e.g., Kwasniok 1996; Achatz and Schmitz 1997), but simpler cases such as *Principal Oscillation Patterns* (von Storch et al. 1995) and regressions techniques are abundant in the literature.

11.4.1 Principal Interaction Patterns[3]

To formalise the PIP concept, we assume that the full system is represented by a state space model (11.1) and a linear *observation equation* (11.2) for the observed variables ω,

$$\omega_t = P\phi_t + \text{noise} = \sum_{j=1}^{k} \vec{p}^j \phi_t + \text{noise}. \tag{11.16}$$

with the row vectors $\vec{p}^j$ forming the matrix P – being the Principal Interaction Patterns, spanning a low-dimensional state space. In the state space equation $\phi_{t+1} = F(\phi_t, \vec{\alpha}, t) + \varepsilon$ the operator F represents a class of models that may be nonlinear in the dynamical variables ϕ_t and depends on a set of free parameters $\vec{\alpha} = (\alpha_1, \alpha_2, ...)$.

Different from the data analysis and model verification problem, the dimension k of the state variable ϕ is considered to be much smaller than the dimension of the observations m. Indeed, k is usually of the order of 20 or much less. Matrix P generally has many more columns (m) than rows (k). The system equations 11.1 therefore describe a dynamical system in a smaller phase space than the space that contains ω_t.

The error-term ε in (11.1) is considered here a noise term. It is often disregarded in nonlinear dynamical analyses. However, disregarding the noise in low-order systems ($k < 20$) usually changes the dynamics of the system significantly, since the low-order system is a closed system without noise (cf. Timmermann 1999; von Storch et al. 1999). However, components of the climate system, such as the tropical troposphere or the thermohaline circulation in the ocean are never closed; they continuously respond to "noise" from other parts of the climate system, hence the noise term. It is doubtful if the fundamental assumption, namely that the low-order system is governed by the same dynamics as the full system, is satisfied when the noise is turned off.

When fitting the state space model from Eq. 11.1, 11.16 to a time series, the following must be specified: the class of models F, the patterns P, the free parameters α and the dimension of the reduced system m. The class of models F, must be selected *a priori* on the basis of physical reasoning. The number m might also be specified *a priori*. The parameters α and the patterns P are fitted simultaneously to a time series by minimising the mean square error $\varepsilon[P; \vec{\alpha}]$ of the approximation of the (discretised) time derivative of the observations ω by the state space model:

$$\varepsilon[P; \vec{\alpha}] = E\left(\left\| \omega(t+1) - P\left(F[\phi(t), \vec{\alpha}, t] - \phi(t) \right) \right\|^2 \right) \tag{11.17}$$

---

[3] Following the presentation in von Storch and Zwiers, 1999, Section 15.5.

11.4.2 Principal Oscillation Pattern Analysis[4]

A standard "cognitive model" is the identification of normal modes or "waves" in geophysical fluid dynamics. In one approach, the basic equations are simplified and linearised until they may be formulated as

$$\omega_{t+1} = A\omega_t + \varepsilon \qquad (11.18)$$

where A is the "system's matrix" and ε the error introduced by the manipulations. The eigenvectors of the matrix A are the normal modes. Typical examples for such modes are all sorts of "waves", as for instance Kelvin waves.

Besides this entirely dynamical approach, there is a statistical variant of Eq. 11.18, namely when then matrix A is *not* derived from dynamical reasoning but fitted to data. In that case ε is usually considered to be *white noise*. Then Eq. 11.18 describes a discrete multivariate first-order autoregressive process.[5] The system matrix A may be estimated through

$$A = \sum\nolimits_1 \sum\nolimits^{-1} \qquad (11.19)$$

where $\sum$ and denote $\sum_1$ the lag-0 and lag-1 covariance matrices of ω, which are easily calculated from the data. The eigenvectors of this estimated matrix A are the *Principal Oscillation Patterns* (POPs). Each state ω may then be expanded, or approximated by the POPs $\vec{e}^k$.

$$\omega = \sum \phi_k \vec{e}^k \qquad (11.20)$$

The ϕ_k are the POP coefficients, and represent the dynamical state of the system:

$$\phi_{k,t+1} = \lambda_k \phi_{k,t} + \varepsilon_k \qquad (11.21)$$

where λ_k is the eigenvalue of A belonging to $\vec{e}^k$. ε_k is the noise projected into the subspace spanned by the POPs.

Note that Eq. 11.20 is an observation equation (11.2) and (11.21) a state equation (11.1).

The POP analysis is illustrated by an application to equatorial variability (von

---

[4] Following von Storch (1993).
[5] The relation between empirical and dynamical modes has been investigated by Schnur et al. (1993), who calculated from quasi-geostrophic theory the dynamical modes describing the extratropical atmospheric variability, and also used the POP approach on a long sequence of analysed geopotential height data. The spatial and temporal characteristics of the most significant POPs were very similar to the most unstable waves in the stability analysis, but the POPs also identified modes representative of the evolution of finite-amplitude waves. Thus, the POPs appear to be useful descriptors of the variability in cases where the dynamics were complex.

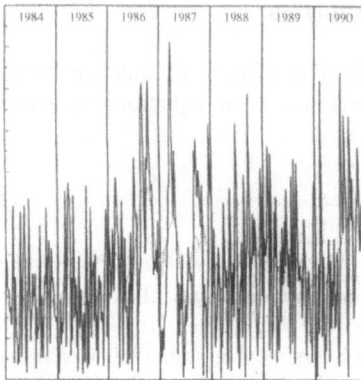 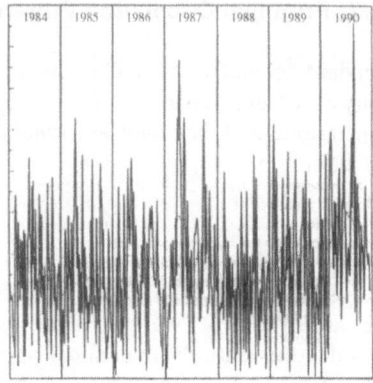

Fig. 11.7. Amplitude time series of two POP modes identified in the daily intraseasonal variability monitored by three Equatorial moorings at 165°E, 140°W and 110°W. The left curve refers to the 120 day mode, and the right one to the 65 day mode. The amplitudes are normalised to standard deviation one. The years are given as May to April intervals (thus "1984" represents the time from May 1984 until April 1985). From von Storch (1993)

Storch 1993). The goal was to investigate the modes of intraseasonal variability in 7 years of moored measurement in the upper equatorial Pacific ocean (Hayes et al. 1991). Oscillatory modes were searched for by using a POP analysis of daily averages of horizontal current at three equatorial locations, 165°E, 140°W and 110°W, for various depths.

The data were first filtered by an EOF analysis to suppress small scale noise, and the EOF coefficients filtered in the time domain to eliminate the variability on periods larger than about half a year. The POP analysis then yielded two oscillatory modes (complex pairs of eigenvectors with the above properties). The normalised amplitude time series are displayed in Fig. 11.7.

One oscillatory mode has a period of $T = 2\pi / \omega = 65$ days and a damping time of $\tau = 73$ days. The amplitude time series reveals an annual cycle with a semi-annual component. The intraseasonal mode activity is strongest during solstice conditions and weakest during equinoctial conditions, and it is enhanced during warm ENSO conditions (1986/87 and 1990).

The other oscillatory mode, operating at a period of about 120 days and a damping time of about 105 days, is affected by the state of the Southern Oscillation as well with enhanced activity during warm episodes and reduced activity during the cold 1988 event.

The spatial amplitudes and phases of the two modes, in terms of zonal currents, are displayed in Fig. 11.8. Both modes represent eastward traveling signals.

The 120-day mode has its largest amplitude, with typical maximum values of about 16 cm/s, at 50 m depth at 65°E and 160 m depth at 140°W. In contrast, the 65 day mode has maximum zonal current anomalies at upper levels (50 m and above) in the eastern part of the basin, with a typical maximum of 12 cm/s at 140°W and 19 cm/s at 110°W. The zonal current 120-day signal propagates in about 60 days from 165°E to 110°W, so that the phase speed is about 1.8 m/s. The

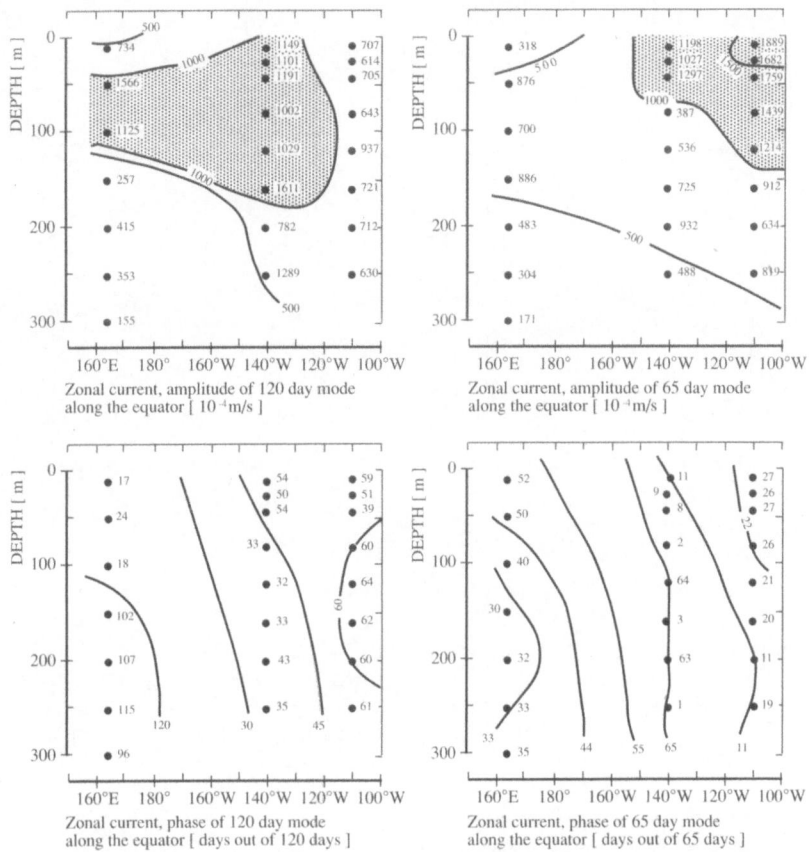

Fig. 11.8. Amplitude and phase distributions of the two eastward propagating oscillatory POPs of the zonal currents at three Equatorial moorings at 165°E, 140°W and 110°W. The coefficient time series are normalised to unity so that the amplitude pattern represents typical distributions in 10^{-4} m/s. The phases are given in days relative to the base period of 120 and 65 days. From von Storch (1993)

phase lines are vertically tilted, with the upper levels lagging the lower levels by about 15 days. The phase speed for the 65 day mode is estimated to be 2.1 m/s. At the two eastern positions, the phase lines are tilted, with the lower levels leading the upper levels by about 8 days. The two modes are not correlated; their time coefficients share a correlation of about −0.25. The two modes are, however, pattern-wise similar and are **not** orthogonal. Indeed, the POP analysis does not require that the modes be orthogonal.

Acknowledgments

Ralf Weisse provided me with the example on the sensitivity of the wave model to the different formulation of the high frequency wind forcing.

References

Abaqus (1998) "Theory manual." Version 5.8. Hibbit, Karlsson and Sorensen, Inc. (8.2.2)

Achatz U and Schmitz G (1997) "On the closure problem in the reduction of complex atmospheric models by PIPs and EOFs: A comparison for the case of a two-layer model with zonally symmetric forcing." J. Atmos. Sci. 54, 2452 – 2474 (11.4)

Achinstein P (1965) "Theoretical models." The British journal for the philosophy of science 16: 102-120 (1.2)

Amani A and Lebel T (1997) "Lagrangian kriging for the Sahelian rainfall at small time steps." Journal of Hydrology, 192, 125 – 157 (10.1)

Arbib MA and Hesse MB (1986) "The construction of reality." Cambridge: Cambridge University Press (1)

Armstrong M and Dowd PA, Eds. (1994) "Geostatistical simulation." Kluwer Academic Publisher, Dordrecht, 255p. (10.5)

Atteia O, Dubois JP and Webster R (1994) "Geostatistical analysis of soil contamination in the Swiss Jura." Environmental Pollution, 86, 315 – 327 (10.2.2)

Backhaus JO (1985) "A three-dimensional model for the simulation of shelf sea dynamics." Deutsche Hydrographische Zeitschrift, 38, 165 – 187 (5.3.5)

Baker JM (1983) "Impact of oil pollution on living resources." Comm. Ecol. Paper No. 4. Interna-tional Union for Conservation of Nature and Natural Resources, Gland, Switzerland (7.3.2)

Baker JM (1995) "Net environmental benefit analysis for oil spill response." Proceedings, 1995 International Oil Spill Conference. American Petroleum Institute, Washington, D.C.: 611 – 614 (7.4)

Barbier E and Pearce D (1990) "Thinking economically about climate change." Energy Policy (1.8)

Bathe KJ (1996) "Finite element procedures", Prentice Hall, New Jersey 07632 (8.2.3)

Bauer E, Stolley M and von Storch H (2000) "On the response of surface waves to accelerating the wind forcing. Global Atmos. Oc. System." (in press) (11.3.2.3)

Bauer E and Weisse R (2000) "Determination of high-frequency wind variability from observations and applications to North Atlantic wave modeling." Journal of Geophysical Research, 105, No. C11, pp 26,179–26,190, Nov. 15, 2000 (11.3.1)

Beckmann A (1988) "On the vertical structure of midlatitude mesoscale instabilities." *J. Phys. Oceanogr.*, **18**, 1354–1371 (4.2.1)

Beckmann A (1995) "Numerical modeling of time-mean flow at isolated seamounts." In: Müller P and Henderson D (Herausgeber). Topographic Effects in the Ocean. Proceedings 'Aha Huliko'a Hawaiian Winter Workshop, 57–66 (4.2.2.7)

Beckmann A (1998) "The representation of bottom boundary layer processes in numerical ocean cirvulation models." In: E.P. Chassignet and J. Verron (Eds.), *Ocean Modeling and Parameterisation*, Kluwer Academic Publishers, 135–154 (4.2.2.6)

Beckmann A, Böning CW, Brügge B and Stammer D (1994) "On the generation and role of eddy variability in the central North Atlantic Ocean." *J. Geophys. Res.*, **99**, 20381–20391 (4)

Beckmann A, Böning CW, Köberle C and Willebrand J (1994) "Effects of increased horizontal resolution in a simulation of the North Atlantic Ocean." *J. Phys. Oceanogr.*, **24**, 326–344 (4)

Beckmann A and Diebels S (1994) "Effects of the horizontal component of the Earth's rotation on wave propagation on an f-plane." *Geophys. Astrophys. Fl. Dyn.*, **76**, 95–119 (4)

Beckmann A and Haidvogel DB (1997) "A numerical simulation of flow at Fieberling Guyot." *J. Geophys. Res.*, J102 5595–5613 (4.2.2.7)

Beckmann A and Käse RH (1989) "Numerical simulation of the movement of a Mediterranean water lens." *Geophys. Res. Lett.*, **16**, 65–68 (4.2.2.4)

Belluck DA, Hull RN, Benjamin SL, French RD and O´Connell RM (1993) "Defining scientific procedural standards for ecological risk assessment." In: GORSUCH JW, Dwyer FJ, Ingersoll CG and LaPoint TW, eds. Environmental Toxicology and Risk Assessment. 2nd ed. American Society for Testing and Materials, Philadelphia, PA: 440 – 450 (7.4)

Bennett AF (1992) "Inverse methods in physical oceanography." Cambridge University Press, Cambridge, 346p (10.4.4)

Bernem KH van (1982) "Effects of experimental crude oil contamination on abundance, mortality and resettlement of representative mud flat organisms in the mesohaline area of the Elbe estuary." Neth. J. of Sea Res., 16: 538 – 546 (7.3)

Bernem KH van (1992) "Thematische Kartierung und Sensitivitätsraster im Deutschen Wattenmeer." Deutsche Hydrogrographische Zeitschrift, 44.Jahrg.1991/92, Heft 5/6: 293 – 309 (7.4)

Bernem KH van, Dörjes J and Müller A (1989) "Environmental oil sensitivity of the German North Sea Coast." Proceedings, 1989 International Oil Spill Conference, American Petroleum Institute, Washington D.C.: 239 – 245 (7.3.2)

Bernem KH van, Müller A, Grotjahn M, Knüpling J, Neugebohrn L, Ramm G, Sach G and Suchrow S (1992) "Thematic mapping and sensitivity study of mud-flat areas in the German Wadden Sea." Neth. J. of Sea Res. 20: 237 – 238 (7.4)

Beucher S (1990) *"Segmentation d' images et morphologie mathématique."* Thèse, École Nationale Supérieure des Mines de Paris, June 1990 (9.3.3)

Beukema JJ (1981) "Quantitative data on the benthos of the Wadden Sea proper." In: Invertebrates of the Wadden Sea, Dankers N, Kühl H and Wolff WJ, eds. Belkema, Rotterdam: 112 – 174 (7.3.2)

Biau G, Zorita E, von Storch H and Wackernagel H (1999) "Estimation of precipitation by kriging in the EOF space of the sea level pressure field." Journal of Climate, 12, 1070 – 1085 (10.1)

Böhme W, Sun D-Z, Schmitt W and Hönig A (1992) "Application of micromechanical material models to the evaluation of Charpy tests." ASME-Symposium: Advances in Local Fracture/Damage Models for the Analysis of Engineering Fracture Problems, AMD-Vol. 137, Nr. H00741 (Edts. GIOVANOLA HS, ROSAKIS AJ), Scottsdale, Arizona, 203 – 216 (8.3.1)

Boyd JP (1985) "Equatorial solitary waves." Part 3: Westward-traveling modons. *J. Phys. Oceangr.,* **15** 46-54 (4.2.2.1)

Boyd JP (1989) "Chebyshev and Fourier spectral methods." Lecture notes in engineering, 49, Springer-Verlag, Berlin (4.2.2.1)

Boyer DL and Zhang X (1990) "Motion of oscillatory currents past isolated topography." *J. Phys. Oceanogr.,* **20** 1425–1448 (4.2.2.7)

Braudel F ([1979] 1992) "The structures of everyday life. The limits of the possible." Volume 1 of civilization and capitalism 15th -18th Century. Berkeley: University of California Press (1.9)

Brink KH (1989) "The effect of stratification on seamount-trapped waves." *Deep-Sea Res.,* **36** 825–844 (4.2.1.1)

Brink KH (1990) "On the generation of seamount-trapped waves." *Deep-Sea Res.,* **3 7** 1569–1582 (4.2.2.7)

Brink KH (1995) "Tidal and lower frequency currents above Fieberling Guyot." *J. Geophys. Res.,* **100** 10817–10832 (4.2.2.7)

Britter R.E. and Mc Quaid J. (1988) "Workbook on the dispersion of dense gases, Health and Safety Executive." Sheffield, U.K. (6.3)

Brocks W, Klingbeil D, Künecke G and Sun D-Z (1995) "Application of the Gurson model to ductile tearing resistance." Second Symposium on Constraint Effects, ASTM STP 1224 (Edts. Kirk M und Bakker A), Philadelphia: American Society for Testing and Materials, 232 – 252 (8.3.1)

Brodbeck M ([1959] 1968) "Models, meaning, and theories." Pp. 579-600 in May Brodbeck (ed.), Readings in the philosophy of the social sciences. New York: Macmillan

Brulin O and Hsieh RKT (Edts.) (1981) "Continuum models of discrete systems IV." Proc. Fourth Int. Conf., Stockholm, June 29 – July 3, 1981, Amsterdam: North-Holland Publ. Comp. (8.3.2)

Bülow KG, Callies U and Rhodin A (1998) "Sensitivity studies using a nutrient model for the River Elbe." In: V. Babovic and L.C. Larsen (Eds.): Hydroinformatics '98, A.A. Balkema, Rotterdam, Brookfield. ISBN 90 5410 983, 1299 – 1304 (11.3.3)

Campbell N R (1920) "Physics, the elements." Cambridge: Cambridge University Press (1.1)

Cartwright N (1983) "How the laws of physics lie." Oxford: Clarendon Press (1.1)

Casti JL (1989) "Alternate realities: mathematical models of nature and man." Wiley, New York, 493 pp (5.2.2)

Champion R, Lenard CT and Mills TM (1996) "An introduction to abstract splines." *Mathematical Scientist*, 21, 8 – 26 (10.4.4)

Chapman DC and Gawarkiewicz G (1995) "Offshore transport of dense water in the presence of a submarine canyon." *J. Geophys. Res.*, **100** 13373-13387 (4.2.2.6)

Chapman DC and Haidvogel DB (1992) "Formation of Taylor caps over a tall isolated seamount in a stratified ocean." *Geophys. Astrophys. Fl. Dyn.*, **64** 31-65 (4.2.2.7)

Chapman DC and Haidvogel DB (1993) "Generation of internal lee waves trapped over a tall isolated seamount." *Geophys. Astrophys. Fl. Dyn.*, **69** 31-65 (4.2.2.7)

Charpy G (1901) "Testing of metals by impact bending of notched bars." Mémoires de la Société des Ingénieurs Civils de France (8.2.1)

Chassignet EP and Bleck R (1993) "The influence of layer outcropping on the separation of boundary currents. Part I. The wind-driven experiments." *J. Phys. Oceanogr.*, **23** 1485-1507 (4.2.2.2)

Chassignet EP, Bleck R and Rooth CGH (1995) "The influence of layer outcropping on the separation of boundary currents." Part II. The wind- and buoyancy-driven experiments. *J. Phys. Oceanogr.*, **25** 2404-2422 (4.2.2.2)

Chauvet P, Pailleux J and Chiles JP (1976) "Analyse objective des champs météorologiques par cokrigeage." La Meteorologie, 6e serie, No 4, 37 – 54 (10.1)

Chiles JP and Delfiner P (1999) "Geostatistics: Modeling spatial uncertainty." Wiley, New York, 695p (10.5)

Codiga DL (1993) "Laboratory realisations of stratified seamount-trapped waves." *J. Phys. Oceanogr.*, **23** 2053–2071 (4.2.2.7)

Collins Compact English Dictionary (1998) (8.1)

Coster M and Chermant JL (1985) *"Précis d'analyse d'Images."* CNRS Ed., Paris (9.4)

Cressie N (1990) "Reply to letter by G. Wahba." The American Statistician, 44, 256 – 258 (10.1)

Cressie N (1991) "Statistics for spatial data." Wiley, New York, 900p. (10.1)

Crowley TJ and North GR (1991) "Paleoclimatology." Oxford University Press, New York, 330 pp (2.3.2)

Dahling PS, Mackay D, Johansen O and Brandvik PJ (1990) "Characterisation of crude oil for environmental purposes." Oil and Chemical Pollution, v7: 199 – 24 (7.4)

Daley R (1991) "Atmospheric data analysis." Cambridge University Press, 457p. (10.1)

Damm PE (1997) "Die saisonale Salzgehalts- und Frischwasserverteilung in der Nordsee und ihre Bilanzierung." (in German), PhD thesis, Berichte aus dem Zentrum für Meeres- und Klimaforschung der Universität Hamburg, Reihe B:Ozeanographie, Nr.28, 259 pp (5.3.7)

Delfine, P (1976) "Linear estimation of non stationary spatial phenomena." In: Guarascio, M. et al. (eds) Advanced Geostatistics in the Mining Industry, 49 – 68, NATO ASI Series C 24, Reidel, Dordrecht (10)

Dengg J (1993) "The problem of Gulf stream separation: A barotropic approach." *J. Phys. Oceanogr.*, **23** 2182–2200 (4.2.2.2)

Dengg J, Beckmann A and Gerdes R (1996) "The Gulf Stream separation problem." In: W. Krauss (Herausgeber), *The Warmwatersphere of the North Atlantic Ocean*. Gebr. Borntraeger, 253-290 (4.2.2.2)

Dicks B and Wright R (1989) "Coastal sensitivity mapping for oil spills. In: Ecological Impacts of the Oil Industry, B. Dicks, ed. 1989 Institute of Petroleum (7.3)

Dörjes J (1984) "Experimentelle Untersuchungen zur Wirkung von Rohöl und Rohöl/Tensid-Gemischen im Ökosystem Wattenmeer." XVI. Zusammenfassung und Schlußfolgerungen. Senckenbergiana marit. 16: 267 – 271 (7.3)

Drucker P (1993) "The ecological vision. Reflections on the American condition." New Brunswick, New Jersey: Transaction Books (1.8)

Dubrule O (1984) "Comparing splines and kriging." Computers and Geosciences, 10, 327 – 338 (10.1)

Duhem P ([1914] 1954) "The aim and structure of physical theory." Princeton, New Jersey: Princeton University Press (1.1)

DYNAMO group [Barnard S, Barnier B, Beckmann A, Böning CW, Coulibaly M, Cuevas DA de, Dengg J, Dieterich Ch, Ernst U, Herrmann P, Jia Y, Killworth PD, Kröger J, Lee MM, LeProvost Ch, Molines JM, New AL, Oschlies A, Reynaud T, West LJ, Willebrand J] (1997) "DYNAMO - Dynamics of North Atlantic models: Simulation and assimilation with high resolution models." *Ber. Inst. f. Meereskunde Kiel*, **294**, 333 pp (4)

Ekman VW (1905) "On the influence of the earth's rotation on ocean currents." *Ark. f. Mat., Astron. och Fysik*, **2**, 1–53 (4.2.1.2)

Elzen M den (1994) "Global environmental change. An integrated modeling approach." Utrecht: International books (1)

Engelen G, White R, Uljee I and Drazan P (1995) "Using cellular automata for integrated modeling of socio-envionmental systems." Environmental Monitoring and Assessment, 34, 203 – 214 (5.2.2)

Engelen G, White R, Uljee I (1997) "Integrating constrained cellular automata models, GIS and decision support tools for urban planning and policy making." In: Decision Support Systems in Urban Planning, H.P.J. Timmermanns (ed.), E&FN Spon, London (5.2.2)

Eringen AC (1967) "Mechanics of continua." John Wiley (8.2.2)

ESDU (1974) Characteristics of atmospheric turbulence near the ground, Part II: Single point data for strong winds (neutral atmosphere), Engineering Science Data Unit, 251 – 259 Regent Street, London W1R 7AD (6.2)

Farke H, Blome D, Theobald N and Wonneberger K (1985) "Field experiments with dispersed oil and a dispersant in an intertidal ecosystem: fate and biological effects." Proceedings, 1985 International Oil Spill Conference, American Petroleum Institute, Washington, D.C.: 515 – 520 (7.3)

Fennel W and Schmidt M (1991) "Responses to topographical forcing." *J. Fluid. Mech.*, **223** 209–240 (4.2.2.6)

Flather RA, Smith JA, Richards JD, Bell C and Blackman DL (1998) "Direct estimates of extreme storm surge elevations from a 40 year numerical model simulation and from observations." Global Atm. Oc. System (in press)

Fohrmann H, Backhaus JO, Blaume F and Rumohr J (1998) "Sediments in bottom arrested gravity plumes - numerical case studies." *J. Phys. Oceanogr.*, **28** 2250–2274 (4.2.3.2)

François D, Pineau A and Zaoui A (1998) "Mechanical behaviour of materials." Vol I: "Elasticity and plasticity", Vol. II: "Viscoplasticity, damage, fracture and contact mechanics", Dordrecht: Kluwer Academic Publ. (8.2.2)

Frankignoul C (1979) "Stochastic forcing models of climate variability." Dyn. Atmos. Oc. 3, 465 – 479 (11.1)

Fredriksson B and Sjöström L (1997) "The role of mechanics and modeling in advanced product development." European Journal of Mechanics, A/Solids, Vol **16** (1997), 83 – 86 (8.1)

Funtowicz SO and Ravetz JR (1990) "Uncertainty and quality in science policy." Dordrecht: Kluwer (1.6)

Galilei G (1638) "Discorsi e demostrazioni matematiche, intorno à due nueve scienze attenti alla mechanica and i movimenti locali." Elsevir (8.2.1)

Gandin, L., 1963: "Objective analysis of meteorological fields." Leningrad, Gidromet (English translation: Israel Program for Scientific Translation, Jerusalem, 1965) (10.1)

Gawarkiewicz G and Chapman DC (1995) "The role of stratification in the formation and maintenance of shelf-break fronts." *J. Phys. Oceanogr., J. Phys.. Oceanogr.*, **22** 753–772 (4.2.2.6)

Gawarkiewicz G and Chapman DC (1995) "A numerical study of dense water formation and transport on a shallow, sloping continental shelf." *J. Geophys. Res.*, **100** 4489–4507 (4.2.2.6)

Giere O (1979) "The impact of oil pollution on intertidal meiofauna." Field studies after the La Coruna spill, May 1976. Cah. Biol. Mar., 20: 231 – 251 (7.3)

Gill AE (1982) "*Atmosphere-ocean dynamics.*" International Geophysics Series, **30**. Academic Press (4.1)

Godard O (1992) "Social decision making in the context of scientific controversy: The interplay of environmental issues, technological conventions and economic stakes." Global Environmental Change 239-249 (1)

Goldberg LP (1989) "Interconnectedness in nature and in science: The case of climate and climate Modeling." Cooperative Thesis No. 121. Boulder, Colorado: University of Colorado and National Center for Atmospheric Research (1)

Gordon S (1991) "The history and philosophy of social science." London and New York: Routledge (1.4)

Goulard M and Voltz M (1992) "Linear coregionalisation model: tools for estimation and choice of multivariate variograms." Mathematical Geology, 24, 269 – 286 (10.3.1)

Grove-White R (1994) "Environmentalism. A new moral discussion for technological society?" Pp. 18-30 in Kay Milton, Kay (ed.), Environmentalism. A view from anthropology. London: Routledge (1)

Grzebyk M and Wackernagel H (1994) "Multivariate analysis and spatial/temporal scales: real and complex models." Proceedings of XVIIth International Biometrics Conference, Volume 1, 19 – 33, Hamilton, Ontario (10.3.4)

Gundlach, E.R. and Hayes MO (1978) "Vulnerability of coastal environments to oil spill impacts." Mar. Technol. Soc. J. 12: 18 – 27 (7.3)

Gurson AL (1977) "Continuum theory of ductile rupture by void nucleation and growth: Part I – Yield criteria and flow rules for porous ductile media." J. Engng. Materials and Technology 99, 2 – 15 (8.3.1)

Gurtin ME (1981) "An introduction to continuum mechanics." Academic Press (8.2.2)

Haidvogel DB and Beckmann A (1998) "Numerical modeling of the coastal ocean." In: Brink, K.H. and A.R. Robinson (Eds.): *The Sea*, **Vol. 10**, 457-482 (4.2.2.7)

Haidvogel DB, Beckmann A and Hedström KS (1991) "Dynamical simulations of filament formation and evolution in the Coastal Transition Zone." *J. Geophys. Res.*, **96**, 15017-15040 (4.2.1.1)

Haidvogel DB, Beckmann A, Chapman DC and Lin R-Q (1993) "Numerical simulation of flow around a tall isolated seamount: Part II: Resonant generation of trapped waves." *J. Phys. Oceanogr.*, **23** 2373-2391 (4.2.2.7)

Haidvogel DB, Wilkin JL and Young RE (1991b) "A semi-spectral primitive equation ocean circulation model using vertical sigma and orthogonal curvilinear horizontal coordinates." *J. Comp. Phys.*, **94** 151-185 (4.2.2.3)

Hall DH, Hollis EJ. and Ishaq H (1982) "A wind tunnel model of the Porton Down dense gas spill field trials." Report LR 394 (AP). Warren Spring Laboratory, Stevenage, U.K. (6.6)

Hansen W (1952) "Gezeiten und Gezeitenströme der halbtägigen Hauptmondtide M^2 in der Nordsee." Deutsche Hydrographische Zeitschrift, Erg.-H. (A) Nr.1, 46 S (5.3.7)

Hasselmann K (1976) "Stochastic climate models. Part I. Theory." Tellus 28, 473-485 (2.1.3)

Hasselmann K (1988) "PIPs and POPs: The reduction of complex dynamical systems using principal interaction and oscillation patterns." J. Geophys. Res. 93, 11015 – 11021 (2.3.2, 11.4)

Havens JA and Spicer TO (1984) "Gravity spreading and air entrainment by heavy gas instantaneously released in a calm atmosphere." In: Atmospheric Dispersion of Heavy Gases and Small Particles (edited by Ooms G. and Tennekes H.). Springer-Verlag, Berlin (6.6)

Hejimans HJAM (1994) "*Morphological Image Operators.*" Academic Press, Boston (9.1)

Heinrich M and Gerhold E (1986) "Praxisgerechte Bestimmung der Zündentfernung bei der Freisetzung schwerer Gase (LPG)." 3. Zwischenbericht, TÜV Norddeutschland (6.8)

Hershner, C and Moore K (1977) "Effects of the Chesapeake Bay oil spill on salt marshes of the lower bay." Proceedings, 1977 International Oil Spill Conference, American Petroleum Institute, Washington, D.C.: 529 – 533 (7.3.2)

Hesse MB (1966) "Models and analogies in science." Notre Dame, Indiana: University of Notre Dame Press (1.2)

van Heugen HH and Duijm NJ (1984) "Some findings based on wind tunnel simulations and model calculations of Thorney Island Trial No. 008." J. Haz. Mat. 11, 409 – 416 (6.6)

Hodge PG (1970) "Continuum mechanics." London: McGraw-Hill Publ. Comp. (8.2.2)

Holloway G (1992) "Representing topographic stress for large-scale ocean models." *J. Phys. Oceanogr.,* **22** 1033-1046 (4.2.2.7)

Honerkamp J (1990) "Stochastic dynamical systems." VCH, Weinheim (3.7)

Honerkamp J (1994) "Stochastic dynamical systems: concepts, numerical methods, data." VCH Publishers, ISBN 3-527-89563-9, 535 pp. (11.3.3)

Huang HC and Cressie N (1996) "Spatio-temporal prediction of snow water equivalent using the Kalman filter." Computational Statistics and Data Analysis, 22, 159 – 175 (10.1)

Hudson G and Wackernagel H (1994) "Mapping temperature using kriging with external drift: theory and an example from Scotland." International Journal of Climatology, 14, 77 – 91 (10.3.5)

Huntington, Ellsworth (1927) "The human habitat." New York: Van Nostrand (1.9)

Huppert HE and Bryan K (1976) "Topographically generated eddies." *Deep-Sea Res.,* **23** 655-679 (4.2.2.7)

Hutchinson MF and Gessler PE (1994) "Splines – more than just a smooth interpolator." Geoderma, 62, 45 – 67 (10.1)

Irwin HPAH (1981) "The design of spires for wind simulation." J. Wind. Eng. Ind. Aerodyn. 7, 361 – 366 (6.2)

Jahn Th. (1990) "Das Problemverständnis sozial-ökologischer Forschung. Umrisse einer kritischen Theorie gesellschaftlicher Naturverhältnisse." Jahrbuch für sozial-ökologische Forschung 1: 15-41

Janssen LAM (1981) "Wind tunnel modeling of the dispersion of LNG vapour clouds in the atmospheric boundary layer." MT-TNO Rep. 81–07020 (6.4)

Jiang L and Garwood RW (1996) "Three-dimensional simulations of overflows on continental slopes." *J. Phys. Oceanogr.,* **26** 1214–1233 (4.2.2.6)

Jones RH (1985) "Time series analysis – time domain." In: A.H. Murphy and R. W. Katz (Eds.): Probability, Statistics, and Decision Making in the Atmospheric Sciences. Westview Press, Boulder and London, ISBN 0-86531-152-8, 223 – 260 (11.3.3)

Jones H and Marshall J (1993) "Convection with rotation in a neutral ocean: a study of open ocean deep convection." *J. Phys. Oceanogr.,* **23** 1009–1039 (4.2.2)

Journel A and Huijbregts C (1978) "Mining geostatistics." Academic Press, London, 600p.

Jungclaus JH and Backhaus JO (1994) "Application of a transient reduced gravity plume model to the Denmark Strait Overflow." *J. Geophys. Res.,* **99** 12375–12396 (4.2.3.2)

Jungelaus JH and Mellor GL (1999) "A three-dimensional model study of the Mediterranean outflow." Submitted to *Journal of Marine Systems* (4.2.2.6)

Kaimal JC, Wyngaard JC, Izumi Y and Cote PR (1972) "Spectral characteristics of surface layer turbulence." Quart. Journ. Roy. Met. Soc., 98, 563 – 588 (6.2)

Kalnay EM, Kanamitsu R, Kistler W, Collins D, Deaven L, Gandin M, Iredell S, Saha G, White J, Woollen YZH, Chelliah M, Ebisuzaki W, Higgins W, Janowiak J, Mo K.C,

Ropelewski C, Wang J, Leetmaa A, Reynolds R, Jenne R and Joseph D (1996) "The NCEP/NCAR 40-year reanalysis project." Bulletin of the American Meteorological Society, Vol. 77, No. 3, 437-471 (2.1.4)

Kates RW (1985) "The interaction of climate and society." Pp. 4-35 in Robert W. Kates, Jesse H. Ausubel and M. Berberian (eds.), Climate Impact Assessment. New York: Wiley (1.1)

Katz RW and Brown BG (1992) "Extreme events in a changing climate: Variability is more important than averages." Climatic Change 21: 289-302

Kauker F (1998) "Regionalisation of climate model results for the North Sea." PhD thesis University of Hamburg, 109 pp. (11.1)

Kellogg WW (1978) "Global influences of mankind on the climate." Pp. 205-227 in John Gibbon (ed.), Climate Change. Cambridge: Cambridge University Press (1 Appendix)

Kielmann J and Käse RH (1987) "Numercial modeling of mneander and eddy formation of the Azores-Current frontal zone." *J. Phys. Oceanogr.*, **17** 529–541 (4.2.2.3)

Kitanidis PK (1997) "Introduction to Geostatistics: Applications to hydrogeology." Cambridge University Press, Cambridge, 249p. (10.1)

Klapp E (1971) "Wiesen und Weiden." 4th ed. Paul Parey, Berlin and Hamburg. 620 pp (7.3.2)

Klein WH and Lahn HR (1974) "Forecasting local weather by means of model output statistics." Bulletin Amer. Met. Soc. 55, 1217 – 1227 (11.3.2.1)

Klinck JM (1996) "Circulation near submarine canyons: a modeling study." *J. Geophys. Res.*, **101** 1211–1223 (4.2.2.7)

König G (1987) "Windkanalmodellierung der Ausbreitung störfallartig freigesetzter Gase schwerer als Luft, Hamburger Geophysikalische Einzelschriften." Verlag Wittenborn Söhne, Hamburg (6.10)

König G, Schatzmann M and Lohmeyer A (1987) "Measurements of gas concentration fluctuations in wind tunnel simulations." Proceedings 7th International Conference on Wind Engineering, Aachen, F.R.G. (6.5)

König-Langlo G and Schatzmann M (1988) "Modeling the dispersion of accidentally-released toxic gases heavier than air." Proc., 17th International Technical Meeting on Air Pollution Modeling and its Application, Cambridge, U.K. (6.10)

Koopmann RP et al. (1982) "Burro Series Data Report, LLNL/NWC 1980 LNG Spill Tests." Lawrence Livermore National Laboratory, U.S.A. (6.8)

Kraus EB and Turner JS (1967) "An one-dimensional model of the seasonal thermocline." II. The general theory and its consequences. *Tellus*, **19** 98–106 (4.2.2.5)

Kunze E and Toole JM (1997) "Tidally driven vorticity, diurnal shear, and turbulence atop Fieberling seamount." *J. Geophys. Res.*, **27** 2663–2693 (4.2.2.7)

Kwasniok F (1996) "The reduction of complex dynamical systems using principal interaction patterns." Physica D 92, 28 – 60 (11.4)

Kyriakidis PC and Journel AG (1999) "Geostatistical space-time models: a review." Mathematical geology, **31**, 651 – 684 (10.5)

Lai WM, Rubin D and Krempl E (1993) "Introduction to continuum mechanics." Pergamon Press (8.2.2)

Landau LD, Lifshitz EM (1960) "Course of theoretical physics, Vol. 1, Mechanics." Pergamon Press, Oxford (3.2)

Landau LD, Lifshitz EM (1960) "Course of theoretical physics, Vol. 6, Fluid mechanics." Pergamon Press, Oxford (3.3)

Landau LD, Lifshitz EM (1960) "Course of theoretical physics, Vol. 5 Statistical physics, Part 1." Pergamon Press, Oxford (3.4)

Landau LD, Lifshitz EM (1960) "Course of theoretical physics, Vol. 9 Statistical physics, Part 2." Pergamon Press, Oxford (3.6)

Langenberg H (1996) "Strange attractors in cellular automata." In: Proceedings of the European Conference on Iteration Theory, Batschuns, Austria, Sept. 9 – 13 1992. Foerg-Rob, Gronau, Mira, Netzer and Targonski (eds.), World Scientific, Singapore: 172 – 181 (5.2.2)

Langenberg H, Pfizenmayer A, von Storch H and Sündermann J (1999) "Storm related sea level variations along the North Sea coast: natural variability and anthropogenic change." – Cont. Shelf Res. **19**, 821 – 842 (11.3.2.2)

Langenberg H and Pohlmann T (1994) "The effect of a front on the general circulation – A model study of the Rhine plume area." Deutsche Hydrographische Zeitschrift, 46, No. 4, 341 – 353 (5.4.3)

Lantuéjoul C and Schmitt M (1991) "Use of two new formulae to estimate the Poisson intensity of a Boolean model." In GRETSI, volume 2, pages 1045 – 1048, Juan-Les-Pins, September 1991

Lantuéjoul Ch (1996) "Iterative algorithms for conditional simulations." In 5th International Geostatistics Congress, pages 27 – 40, Wollongong, Australia, September 22 – 27 (9.5)

Lantuéjoul Ch (1999) "*Géostatistical simulation:* Models and algorithms." Springer (9.5)

Large W (1998) "Modeling and parameterisation ocean planetary boundary layers." In: E.P. Chassignet and J. Verron (Eds.), *Ocean Modeling and Parameterisation*, Kluwer Academic Publishers, 81–120 (4.2.2.5)

Large WG, McWilliams JC and Doney SC (1994) "Oceanic vertical mixing: a review and a model with a nonlocal boundary layer parameterisation." *Rev. Geophys.*, **32**, 363–403

LeBlond PH and Mysak LA (1978) "Waves in the ocean." Elsevier Oceanography Series, **20**, Elsevier, Amsterdam (4B)

Le Cun Y (1987) "*Modèles Connexionnistes de l'Apprentissage.*" PhD thesis, Université Pierre et Marie Curie, Paris, France (9.3.5)

Lemaitre J and Chaboche J-L (1990) "Mechanics of solid materials." Cambridge: University Press (8.2.2)

Le Roy Ladurie E ([1967] 1988) "Times of feast, times of famine: A history of climate since the year 1000." New York: Farrar, Strauss and Giroux (1.9)

Lindstedt-Siva J (1991) "U.S. oil spill policy hampers response and hurts science." Proceedings, 1991, International Oil Spill Conference. American Petroleum Institute, Washington, D.C.: 349 – 352 (7.4)

Liverman DM (1983) "The use of a simulation model in assessing the impacts of climate on the world food system." University of California, Los Angeles-National Center for Atmospheric Research Cooperative Thesis No. 77. Boulder, Colorado

Livezey RE (1995) "The evaluation of forecasts." In: H. von Storch and A. Navarra (Eds) Analysis of climate variability: Applications of statistical techniques, Springer Verlag, 177-196 (ISBN 3-540-58918-X) (2.3)

Loder JW (1980) "Topographic rectification of tidal currents on the Sides of Georges Bank." *American Meteorological Society*, **10**, 1399-1416

Lohmann U and Roeckner E (1995) "Influence of cirrus cloud radiative forcing on climate and climate variability in a general circulation model. J. Geophys. Res. 100 D, 16305 – 16323 (11.3.2.3)

Lorenz EN (1963) "Deterministic nonperiodic flow." J. Atmos. Sci. 20, 130-141 (2.1.3)

Lundvall B-Å (1992) "Introduction." Pp. 1-19 in Bengt-Åke Lundvall (ed.), National systems of innovation. Aalborg: Aalborg University Press (1.2)

Luyten JR, Pedlosky J and Stommel H (1983) "The ventialted thermocline.". *J. Phys. Oceanogr.*, **13** 292-309 (4.2.1.3)

Luyten JR and Stommel H (1986) "Gyres driven by combined wind and bupyancy flux." *J. Phys. Oceanogr.*, **16** 1551-1560 (4.2.1.4)

Maas LRM and Zimmerman JTF (1989a) "Tide-topography interactions in a stratified shelf sea I. Basic equations for quasi-nonlinear internal tides." *Geophys. Astrophys. Fl. Dyn.*, **45** 1-35 (4.2.2.7)

Maas LRM and Zimmerman JTF (1989b) "Tide-topography interactions in a stratified shelf sea II. Bottom trapped internal tides and baroclinic residual currents." *Geophys. Astrophys. Fl. Dyn.*, **45** 37-69 (4.2.2.7)

Mackay A (1981) "Climate and popular unrest in late medieval Castile." Pp. 356-376 in T.M.L. Wigley, M.J. Ingram and G. Farmer (eds.), Climate and History. Studies in Past Climates and their Impact on Man. Cambridge: Cambridge University Press (1 Appendix)

Maier-Reimer E, Mikolajewicz U and Crowley T (1990) "Ocean general circulation model sensitivity experiments with an open Central America isthmus." Paleooceanography 5, 349 – 366 (11.3.2.3)

Maisonneuve F (1982) "Sur le partage des eaux." Technical report, CGMM, École des Mines de Paris (9.1)

Malenson GP (1993) "Comment on modeling ecological responses to climatic change." Climatic Change 23: 95-109

Malone Th and Gafy Y (1992) "Towards a general method for analysing regional impacts of global change." Global Environmental Change 101-110 (1 Appendix)

Mannheim K (1929) "Ideologie und Utopia." Bonn: Cohen (1.1)

Mardia KV, Kent JT, Goodall CR and Little JA (1996) "Kriging and splines with derivative information. Biometrika, 83, 207 – 221 (10.1)

Marotzke K. (1991) Spectra of longitudinal turbulence measured in a boundary layer wind tunnel, Atmospheric Enviroment, 24 A, 1277 – 1282 (6.2)

Marotzke K und Schatzmann M (1993) Untersuchung von Hindernisstrukturen bei der Störfallausbreitung, Forschungsbericht UBA 104 09 110, Meteorologisches Institut, Universität Hamburg (6.10)

Martin B (1988) "Nuclear winter: Science and politics." Science and Public Policy 15: 321-334

Matheron G (1967) *"Éléments pour une théorie des milieux poreux."* Masson, Paris (9.1)

Matheron G (1973) "The intrinsic random functions and their applications." Advances in Applied Probability, 5, 439 – 468 (10.1)

Matheron G (1975) *"Random Sets and Integral Geometry."* John Wiley and Sons, New York (9.1)

Matheron G (1979) "Recherche de simplification dans un probleme de cokrigeage." Publication N-628, Centre de Geostatistique, Ecole des Mines de Paris, Fontainebleau (10.2)

Matheron G (1981) "Splines and kriging: their formal equivalence." In: Merriam DF (ed) Down-to-Earth Statistics: Solutions Looking for Geological Problems, 77 – 95, Syracuse University Geology Contribution, No 8, New York (10.4.4)

Matheron G (1983) "Filters and lattices." Technical Report 851, CGMM, Ecole des Mines, September 1983 (9.1)

Matheron G (1986) "Le covariogramme géométrique des compacts convexes de R2." Technical Report 2/86, Centre de Géostatistique, Ecole des Mines de Paris, February 1986 (9.3.2)

Matheron G (1989) "Estimating and choosing." Springer-Verlag, Berlin, 141p. (10.2)

Mattioli J and Schmitt M (1992) "Inverse problems for granulometries by erosion." *Journal of Mathematical Imaging and Vision*, 2:217 – 232 (9.4.1)

Mattioli J (1996) "A textural analysis by mathematical morphology." In R. Schafer P. Maragos and M. But, editors, *Mathematical morphology and ist applications to image and signal processing*, pages 297 – 304. Kluwer Academic Publishers (9.3.2)

Mazur A and Jinling L (1993) "Sounding the global alarm: Environmental issues in the US national news." Social Studies of Science 23: 681-720 (1 Appendix)

McCartney M (1975) "Inertial Taylor Columns on a beta-plane." *J. Fluid. Mech.*, **68** 71-95 (4.2.2.7)

McQuaid J (1984) Large scale experiments on the dispersion of heavy gas clouds, In: Atmospheric Dispersion of Heavy Gases and Small Particles (edited by Ooms G. and Tennekes H.). Springer-Verlag, Berlin (6.3)

McWilliams JC (1985) "Submesoscale coherent vortices in the ocean." *Rev. Geophys.*, **23**, 165-182 (4.2.2.4)

McWilliams JC and Gent PR (1986) "The evolution of submesoscale, coherent vortices on the beta-plane." *Geophys. Astrophys. Fl. Dyn.*, **35** 235-255 (4.2.2.4)

Meadows P (1957) "Models, systems and science." American Sociological Review 22: 3–9

Meincke J (1971) "Observation of an anticyclone vortex trapped above a seamount." *J. Geophys. Res.*, **76** 7432-7440

Mellor GL, Mechoso CR and Keto E (1982) "A diagnostic calculation of the general circulation of the Atlantic Ocean." *Rev. Geophys.*, **29** 1171-1192 (4.2.2.2)

Mellor GL and Yamada T (1982) "Development of a turbulent closure model for geophysical fluid problems." *Rev. Geophys.*, **20**, 851-875 (4.2.2.2)

Mendelsohn R, Nordhaus WD, and Shaw D (1994) "The impact of global warming on agriculture: A Ricardian analysis." American Economic Review: 753–771 (1 Appendix)

Meyer-Abich KM (1980) "Chalk on the white wall? On the transformation of climatological facts into political facts." Pp. 61-92 in Jesse Ausubel and Asit K. Biswas (eds.), Climatic constraints and human activities. Task Force on the Nature of Climate and Society Research, February 4-6, 1980. Oxford: Pergamon Press (1)

Mikolajewicz U and Maier-Reimer E (1990) "Internal secular variability in an OGCM." Climate Dyn. 4, 145-156 (2.2)

Miller A, Mitzer I, and Brown P (1990) "Rethinking the economics of global warming." Issues in Science and Technology Fall (1)

Munk W (1950) "On the wind-driven ocean circulation." *J. Meteorol.*, **7**, 79-93 (4.2.1.3)

Murphy AH and Epstein ES (1989) "Skill scores and correlation coefficients in model verification." Mon. Wea. Rev. 117, 572 – 581 (11.3.2.1)

Najman L and Schmitt M (1993) "Definition and some properties of the watershed of a continuous function." In *Mathematical Morphology and its applications to Signal Processing*, pages 76 – 81, Barcelona, Spain, May, 12 – 14 (9.3.3)

Navarra A (1995) "The development of climate research." In: H. von Storch and A. Navarra (Eds.) "Analysis of climate variability: Applications of statistical techniques", Springer Verlag 3-10 (ISBN 3-540-58918-X) (2.1.4)

Neff DE and Meroney RN (1982) "The behavior of LNG vapor clouds." Report, Contr. No. 5014-352-0203, Gas Research Institute, Chicago (6.6)**Jiang L and Garwood RW (1995)** "A numerical study of three-dimensional dense water bottom plumes on a Southern Ocean continental slope." *J. Geophys. Res.*, **100** 18471–18488 (4.2.2.6)

Nemat-Nasser S. and Hori M (1993) "Micromechanics: overall properties of heterogenous materials." Amsterdam: Elsevier Scientific Publ. (8.3.2)

Needleman A and Tvergaard V (1984) "An analysis of ductile rupture in notched bars." J. Mech. Phys. Solids 32, 461 – 490 (8.3.1)

Nordhaus W (1991) "To slow or not to slow: The economics of the greenhouse effect." The Economist 101: 920-937 (1.8)

Ogburn WF (1946) "The social effects of aviation." Boston: Houghton Mifflin

Ogburn WF (1934) "Studies in the prediction and distortion of reality." Social Forces 13: 227-228

O'Neill RV (1973) "Error analysis of ecological models." Decidous Forest Biome. Memo. Report: 71 – 75 (7.1)

Ooms G and Tennekes H (1984) "Atmospheric Dispersion of Heavy Gases and Small Particles." Springer-Verlag, Berlin

Oreskes N, Shrader-Frechette K and Beltz K (1994) "Verification, validation, and confirmation of numerical models in earth sciences." Science 263, 641-646 (1,6 2.1.4)

Ou HW (1991) "Some effects of a seamount on oceanic flow." *J. Phys. Oceanogr.*, **21** 1835-1845 (4.2.2.7)

Parsons AT (1969) "A two-layer model of Gulf Stream separation." *J. Fluid. Mech.*, **39** 511-528 (4.2.2.2)

Pedlosky J (1987) "Geophysical fluid dynamics." Springer, 710pp (2.3.2)

Pedlosky J (1996) "Ocean circulation theory." Springer-Verlag, 453 pp (4.2.1.4)

Perrings Ch (1987) "Economy and the environment: A theoretical essay on the interdependence of economic and environmental settings." Cambridge: Cambride University Press (1)

Pohlmann Th (1996) "Predicting the thermocline in a circulation model of the North Sea – Part I: Model description, calibration and verification." Cont. Shelf Res., **16**, No. 2, 131 – 146 (5.3.8)

Préteux F and Schmitt M (1988) "Boolean texture analysis and synthesis." In J. Serra, editor, *Image Analysis and Mathematical Morphology, Volume 2:* Theoretical Advances. Academic Press, London

Puttock JS and Colenbrander GW (1985) "Dense gas dispersion – experimental research." In Proceedings of the Heavy Gas (LNG/LPG) Workshop, Toronto, Ontario, Canada (6.8)

Puttock JS, Colenbrander GW and Blackmore DR (1982) "Maplin sands experiments 1980: Dispersion results from continuous releases of refrigerated liquid propane." In Heavy Gas and Risk Assessment – II (edited by Hartwig S.). Reidel Dordrecht, Holland (6.8)

Ramsden D (1995a) "Response of an oceanic bottom boundary layer on a slope to interior flow. Part I: Time-independent interior flow." *J. Phys. Oceanogr.*, **25** 1672-1687 (4.2.2.7)

Ramsden D (1995b) "Response of an oceanic bottom boundary layer on a slope to interior flow. Part II: Time-dependent interior flow." *J. Phys. Oceanogr.*, **25** 1688–1695 (4.2.2.7)

Rees WE and Wackernagel M (1994) "Ecological footprints and appropriated carrying capacity: measuring the natural capital requirements of the human ecology." Pp. 362-390 in AnnMari Jansson, Monica Hammer, Carl Folke and Robert Constanza (eds.), Investing in Natural Capital. The Ecological Economics Approach to Sustainability. Washington, D.C.: Island Press (1.8)

Rivoirard J (1994) "Introduction to disjunctive kriging and non-linear geostatistics." Oxford University Press, Oxford, 181p. (10.5)

Robinson JB (1991) "Modeling the interactions between human and natural systems." International Social Science Journal 130: 629-647 (1)

Robinson AR, Lermusiaux PFJ and Sloan NQ III (1998) "Data assimilation." In: K.H. Brink, A.R. Robinson (eds): The Global Coastal Ocean. Processes and Methods. The Sea Vol. 10. John Wiley & Sons, 541 – 593 (2.3.1, 11.3.3)

Roden GI (1987) "Effect of seamounts and seamount chains on ocean circulation and thermohaline structure." In: Keating B, Fryer P, Batiza R and Bohlert G (Eds.), *Seamounts, Islands and Atolls* Geophysical Monograph **43** American geophysical Union, Washington, D.C. 335–354 (4.2.2.7)

Rodi W (1980) "Turbulence models and their application in hydraulics – a state of the art review." International Association for Hydraulics, Delft (5.3.8)

Ruske W (1971) "100 Jahre Materialprüfung in Berlin – Ein Beitrag zur Technikgeschichte." Berlin: Bundesanstalt für Materialpüfung (BAM) (8.2.1)

Salembier Ph and Serra J (1995) "Flat zones filtering, connected operators and filters by reconstruction." *IEEE Transactions on Image Processing*, 3(8): 1153 – 1160 (9.3.3)

Salembier Ph, Torres L, Meyer F and Gu C (1995) "Region-based video coding using mathematical morphology." *Proceedings of the IEEE*, 83(6): 843 – 857 (9.1)

Sander J, Wolf-Gladrow D and Olbers D (1995) "Numerical studies of open ocean deep convection." *J. Geophys. Res.*, **100** 20579–20600 (4.2.2)

Schär C and Davies H.C (1988) "Quasi-geostrophic stratified flow over isolated finite amplitude topography." *Dyn. Atmos. Oceans.*, **11** 287–306 (4.2.2.6)

Schatzmann M, Donat J, Hendel S and Krishan G (1995) "Design of a low-cost stratified boundary-layer wind tunnel." Journ. Wind Eng. and Aerodyn. 54/55, pp. 483 – 491 (6.2)

Schatzmann M, Rafailidis S, and Duijm NJ (1998) "Wind tunnel experiments. In: Fenger et al. (Ed.) Urban Air Pollution – European Aspects, Kluwer Acad. Publ., Doordrecht, ISBN 0-7923-5502-4 (6.1)

Schelling ThC (1983) "Climatic change: Implications for welfare and policy." Pp. 449-482 in National Research Council, Changing Climate. Report of the Carbon Dioxide Assessment Committee. Washington, D.C.: National Academy Press (1 Appendix)

Schmitt M (1991) "Estimation of the density in a stationary boolean model." *J. of Applied Probability*, 28, September 1991 (9.3.5)

Schmitt M (September 1991) "Variations on a theme in binary mathematical morphology." *J. of Visual Communication and Image Representation*, 2 (3):244 – 258 (9.3.5)

Schmitt M and Beucher H (1996) "Estimation of intensity and shape in a non-stationary boolean model." In *5th International Geostatistics Congress*, Wollongong, Australia, September 22 – 27 (9.5)

Schnur R, Schmitz G, Grieger N and v. Storch H (1993) "Normal modes of the atmosphere as estimated by principal oscillation patterns and derived from quasi-geostrophic theory." J. Atmos. Sci., **50**, 2386 – 2400 (11.4.2)

Schon DA (1963) "The displacement of concepts." London: Tavistock (1.1)

Schott F, Visbeck M. and Send U (1994) "Open-ocean deep convection, Mediterranean and Greeenland Seas." In: P. Malanotte-Rizzoli and A. Robinson (Eds.): *Ocean Processes on Climate Dynamics*. Kluwer, 203–225 (4.2.2)

Schröter J and Wunsch C (1986) "Solution of nonlinear finite difference ocean models by optimisation methods with sensitivity and observational strategy analysis." J. Phys. Oceanogr. **16**, 1855 – 187 (11.3.3)

Schünemann M and Kühl H (1993) "Experimental investigations of the erosional behaviour of naturally formed mud from the Elbe estuary and the adjacent Wadden Sea, Germany." In A. Mehta (ed): Coastal and Estuarine Studies, American Geophysical Union, Washington DC, p. 314-330 (2.1.1)

Schuster HG (1989) "Deterministic Chaos." VCH, Weinheim (3.5)

Seguret S and Huchon P (1990) "Trigonometric kriging: a new method for removing the diurnal variation from geomagnetic data." Journal of Geophysical Research, 32, B13, 21.383 – 21.397

Send U and Marshall J (1995) "Integral effects of deep convection." *J. Phys. Oceanogr.*, **25** 855–872 (4.2.2)

Send U and Käse R (1998) "Parameterisation of proceses in deep convection regimes." In: E.P. Chassignet and J. Verron (Eds.), *Ocean Modeling and Parameterisation*, Kluwer Academic Publishers, 191–214 (4.2.2)

Serra J (1982) *"Image analysis and mathematical morphology.* Academic Press, London (9.1)

Serra J, editor (1988) *Image Analysis and Mathematical Morphology*, Volume 2: Theoretical Advances. Academic Press, London (9.1)

Serra J and Vincent L (January 1992) "An overview of morphological filtering." *Circuits, Systems and Signal Processing*, 11(1): 47 – 108 (9)

Shine KP and Henderson-Sellers A (1983) "Modeling climate and the nature of climate models: A review." Royal Meteorological Society 381-94 (1.6)

Smith PC (1975) "A streamtube model for bottom boundary currents in the oceans." *Deep-Sea Res.*, **22** 853–873 (4.2.2.6)

Stammer D and Böning CW (1992) "Mesoscale variability in the Atlantic Ocean from GEOSAT altimetry and WOCE high resolution numerical modeling." *J. Phys. Oceanogr.*, **22** 732–752 (4.2.1.5)

Starfield AM and Bleloch AL (1983) "Expert system: an approach to problems in ecological management that are difficult to quantify." J. Environ. Manag. 16: 261 – 268 (7.1)

Starfield AM and. Bleloch AL (1986) "Building models for conservation and wildlife management." Collier Macmillan, London (7.1)

Steglich D and Brocks W (1997) "Micromechanical Modeling of the behaviour of ductile materials including particles." Computational Materials Science 9, 7 – 17 (8.3.2)

Stehr N (1992) "Practical knowledge." London: Sage (1.4)

Stehr N (1994) "Knowledge societies." London: Sage (1.6)

Stehr N and von Storch H (1995) "The social construct of climate and climate change." Climate Research 5: 99-105 (1)

Steinberg SR (1986) "Gray scale morphology." *Computer Vision, Graphics and Image Processing*, Vol. 35:333 – 355 (9.1)

Stommel H (1948) "The westward intensification of wind-driven ocean currents." *Trans. Amer. Geophys. Un.*, **29**, 202–206 (4.2.1.3)

Stoyan D, Kendall WS and Mecke J (1987) "*Stochastic geometry and its applications.*" John Wiley and Sons, New-York (9.5)

Streeten P (1972) "Cost-benefit and other problems of method." Pp. 47-59 in Political Economy of Environment: Problems of Method. The Hague: Mouton (1)

Sündermann J and Vollmers H (1972) "Tidewellen in Ästuarien." Wasserwirtschaft 62, 1 – 9 (2.1, 11.1)

Taylor GI (1917) "Motions of solids in fluids when the flow is not irrotational." *Proc. Roy. Soc. Lond. A.,* **93**, 99–113 (4.2.2.6)

Taylor PJ (1995) "Building on construction: an exploration of heterogeneous constructionism, using an analogy from psychology and a sketch from socioeconomic modeling." Perspectives on Science 3:66-98 (1)

Teunissen HW (1982) Validation of boundary layer simulation, Some comparisons between model and full-scale flow, Proc. Int. Workshop on Wind Tunnel Model. Crit. and Techn. for Civ. Eng. Appl., Gaithersburg, Md., USA, Apr. 14 – 16 (6.2)

Thiebaux HJ (1997) "The power of duality in spatial-temporal estimation." Journal of Climate, 10, 567 – 573 (10.1)

Thiebaux HJ and Pedder MA (1987) "Spatial objective analysis: with applications in atmospheric science." Academic Press, London, 299p. (10.1)

Timmermann, A., 1999: Noise-induced transitions in a simplified climate model: A paradigm for abrupt climate change." J. Phys. Oceano. (submitted) (11.3.1)

Trowbridge JH and Lentz SJ (1991) "Asymmetric behavior of an oceanic boundary layer above sloping bottom." *J. Phys. Oceanogr.,* 21 1171–1185 (4.2.2.7)

Truesdell C and Noll W (1965) "The non-linear field theories of mechanics." Encyclopedia of Physics (ed. by S. Flügge), Vol. III/3, Berlin: Springer Verlag (8.2.2)

TÜV Norddeutschland (1982) Forschungsbericht zur Anfertigung einer Störfallstatistik im Rahmen des F&E-Vorhabens "Untersuchung zur Ausbreitung von Luftverunreinigungen im Störfal" (Interner Bericht) (6.8)

Tvergaard V (1982) "On localisation in ductile materials containing spherical voids." Int. J. Fracture 18, 237 – 252 (8.3.1)

Tvergaard V and Needleman A (1984) "Analysis of the cup-cone fracture in a round tensile bar. Acta metall. <u>32</u>, 157 – 169 (8.3.1)

Umweltbundesamt (1983) "Handbuch Störfälle. Dokumentation über Störfälle in Chemieanlagen." Forschungsbericht 104 09 303 UBA-FB-83-023, Umweltbundesamt Berlin (6.6)

United States National Academy of Sceinces (1979) "Carbo dioxide and climate: A Scientific Asessment." Washington, DC: National Academy of Sciences (1.9)

US Environmental Protection Agency (1992a) "Framework for Ecological Risk Assessment." US EPA Document: EPA/630/R-92/001. US Environmental Protection Agency, Washington, D.C. (7.4)

van der Sluijs J, van Eijndhoven J, Shackley S and Wynne B (1998) "Anchoring devices in science and policy: the case of consensus around climate sensitivity." Social Studies of Science 28: 291-323 (1.6)

VDI (1990) "Guideline 3783, Part 2, Dispersion of heavy gas emissions by accidental releases – Safety study." Beuth Verlag, Berlin (6.10)

Vincent L (May 1990) "*Algorithmes morphologiques à base de files d'attente et de locets: Extension aux Graphes.*" Thèse, École des Mines, Paris, France (9.1)

Vincent L (1991) "New trends in morphological algorithms." In SPIE 91, *Nonlinear Image Processing*, volume 1451, pages 158 – 169 (9.1)

von Storch H (1993) "Principal oscillation pattern analysis of the intraseasonal variability in the equatorial Pacific Ocean." Aha' Huliko'a Workshop, 201 – 227 (11.4.2)

von Storch H (1997) "Conditional statistical models: A discourse about the local scale in climate modeling." In P. Müller and D. Henderson (Eds): Monte Carlo Simulations in Oceanography Proceedings 'Aha Huliko'a Hawaiian Winter Workshop, University of Hawaii at Manoa, January 14 – 17, 1997, 59 – 58 (11.3.1)

von Storch H (1999) "The global and regional climate system." In: von Storch H and Flöser G: Anthropogenic Climate Change, Springer Verlag, ISBN 3-540-65033-4, 3-36 (2.1.3)

von Storch H (2000) "Models between academia and applications." In: H. von Storch and G. Flöser (Eds): Models in Environmental Research. Springer Verlag (this volume) (11.1)

von Storch H, Bürger G, Schnur R and von Storch J (1995) "Principal oscillation pattern: A review." – J. Climate 8, 377 – 400 (11.4)

von Storch H, Güss S und Heimann M (1999) "Das Klimasystem und seine Modellierung. Eine Einführung." Springer Verlag (11.4.1)

von Storch H and Hasselmann K (1996) "Climate variability and change." In: Hempel G (ed.): The ocean and the poles. Grand challenges for European cooperation. Gustav Fischer Verlag Jena, Stuttgart, New York, 379 pp., 33-58 (2.1.4)

von Storch H and Zwiers FW (1999) "Statistical analysis in climate research." Cambridge University Press ISBN 0 521 45071 3, 496pp. (2.3.2, 11.3.1)

Wackernagel H (1994) "Cokriging versus kriging in regionalised multivariate data analysis." Geoderma, 62, 83 – 92 (10.3.3)

Wackernagel H (1998) "Multivariate geostatistics: an introduction with applications." 2nd edition, Springer-Verlag, Berlin, 291p. (10.1)

Wahba G (1990a) "Letter to the editor." The American Statistician, 44, 255 – 256 (10.1)

Wahba G (1990b) "Spline models for observational data." Society for Industrial and Applied Mathematics, Philadelphia, 169p. (10.1)

Weber M ([1921] 1972) "Wirtschaft und Gesellschaft." Fünfte revidierte Auflage. Tübingen: J.C.B. Mohr (Paul Siebeck) (1.2)

Weingart P and Maasen S (1997) "The order of meaning: the career of chaos as a metaphor." Configuration 5: 463-52 (1.1)

Wigley TML, Huckstep NJ, Ogilvie AEJ, Farmer G, Mortimer R, and Ingram MJ (1985) "Historical climate impact assessment." Pp. 529-563 in Kates RW, Ausubel J and Berberian M (eds.), Climate Impact Assessment. New York: Wiley (1. Appendix)

Wissel CW (1989) "Theoretische Ökologie – eine Einführung." Springer Verlag, ISBN 3-540-50848-1 (7.1)

Woods JD and Barkmann W (1986) "The response of the upper ocean to solar heating. I. The mixed layer." *Quart. J. Roy. Meteor. Soc.*, **112**, 1–27 (4.2.2.5)

Wright DG and Loder J.W (1985) "A depth-dependent study of the topographic rectification of tidal currents." *Geophys. Astrophys. Fl. Dyn.*, **31** 169–220 (4.2.2.7)

Wynne B (1994) "Scientific knowledge and the global environment." Pp. 169-189 in Michael Redclift and Ted Benton (eds.), Social theory and the global environment. London: Routledge

Zhang X and Boyer DL (1993) "Laboratory study of rotating, stratified, oscillatory flow over a seamount." *J. Phys. Oceanogr.*, **23** 1122–1141

Zienkiewicz OC (1971) "The finite element method in engineering science." London: MacGraw-Hill Publ. Comp. (8.2.3)

Subject Index